Alexander Agassiz

A contribution to American thalassography

Three cruises of the United States Coast and geodetic survey steameer

Alexander Agassiz

A contribution to American thalassography
Three cruises of the United States Coast and geodetic survey steameer

ISBN/EAN: 9783743467125

Manufactured in Europe, USA, Canada, Australia, Japa

Cover: Foto ©berggeist007 / pixelio.de

Manufactured and distributed by brebook publishing software (www.brebook.com)

Alexander Agassiz

A contribution to American thalassography

A Contribution to American Thalassography.

THREE CRUISES

OF THE

UNITED STATES COAST AND GEODETIC SURVEY

STEAMER "BLAKE"

IN THE GULF OF MEXICO, IN THE CARIBBEAN SEA, AND ALONG
THE ATLANTIC COAST OF THE UNITED STATES,
FROM 1877 TO 1880.

BY

ALEXANDER AGASSIZ.

IN TWO VOLUMES.

VOL. II.

BOSTON AND NEW YORK:
HOUGHTON, MIFFLIN AND COMPANY.
The Riverside Press, Cambridge.
1888.

Copyright, 1888,
BY ALEXANDER AGASSIZ.

All rights reserved.

[Published by permission of CARLILE P. PATTERSON and JULIUS E. HILGARD,
Superintendents U. S. Coast and Geodetic Survey.]

The Riverside Press, Cambridge:
Electrotyped and Printed by H. O. Houghton & Co.

CONTENTS OF VOLUME II.

		PAGE
XIV.	THE WEST INDIAN FAUNA	1
XV.	SKETCHES OF THE CHARACTERISTIC DEEP-SEA TYPES. — FISHES. (Figs. 195–224.)	21
XVI.	CHARACTERISTIC DEEP-SEA TYPES. — CRUSTACEA. (Figs. 225–259.)	37
XVII.	CHARACTERISTIC DEEP-SEA TYPES. — WORMS. (Figs. 260–273.)	52
XVIII.	CHARACTERISTIC DEEP-SEA TYPES. — MOLLUSKS.	58
	CEPHALOPODS. (Figs. 274–281)	58
	GASTEROPODS AND LAMELLIBRANCHS. (Figs. 282–312.)	62
	BRACHIOPODS. (Figs. 313–322.)	75
	ASCIDIANS	77
	BRYOZOA. (Figs. 323–337.)	78
XIX.	CHARACTERISTIC DEEP-SEA TYPES. — ECHINODERMS.	84
	HOLOTHURIANS. (Figs. 338–347.)	84
	SEA-URCHINS. (Figs. 348–376.)	88
	STARFISHES. (Figs. 377–387.)	102
	OPHIURANS. (Figs. 388–403.)	109
	CRINOIDS. (Figs. 404–421.)	116
XX.	CHARACTERISTIC DEEP-SEA TYPES. — ACALEPHS.	128
	CTENOPHORÆ AND HYDROMEDUSÆ. (Figs. 422–440.)	128
	HYDROCORALLINÆ. (Figs. 441–448.)	138
XXI.	CHARACTERISTIC DEEP-SEA TYPES. — POLYPS	142
	HALCYONOIDS AND ACTINOIDS. (Figs. 449–461.)	142
	CORALS. (Figs. 462–483.)	148
XXII.	CHARACTERISTIC DEEP-SEA TYPES. — RHIZOPODS. (Figs. 484–519.)	157
XXIII.	CHARACTERISTIC DEEP-SEA TYPES. — SPONGES. (Figs. 520–545.)	170
LIST OF FIGURES		181
INDEX		195

THREE CRUISES OF THE "BLAKE."

XIV.

THE WEST INDIAN FAUNA.

THE inhabitants of the abyssal realm as now known differ far more from the surface faunæ than the latter do from one another, one of the most striking characteristics of deep-sea life being the fact that there exists at the bottom of the ocean a fauna of almost exclusively animal feeders, which, in addition to preying upon one another, receive some of their food from the organic matter living on or near the surface of the sea and constantly falling to the bottom in a decaying condition. The deep-sea fishes, the mollusks, crustacea, and other groups, are nearly all carnivorous, no algæ being found growing at any depth.

Deep-sea forms are almost always killed in the process of hauling, either by rough handling or else by the heat of the surface water. We can scarcely hope ever to watch the habits of the deep-sea dwellers, and see them in their natural attitudes, and we must be satisfied to imagine what these are by analogy with their shallow-water allies, though many species of crustacea, echinoderms, polyps, and mollusks have been kept alive in a casing of ice by the naturalists of the United States Fish Commission. A similar attempt had been made on the "Blake" with some of the echinoderms, but they refused to be deluded for more than a few minutes by ice-cold water into the belief that they still lived in their normal condition.

Very frail deep-sea animals are often rapidly transferred to the surface from a region where they are subjected to a pressure of two tons or more, and it is not surprising that, after having been thus drawn up from a depth of two or three miles, they should be in a very dilapidated condition. A number of the abyssal types among the fishes, mollusks, crustacea, echinoderms, and even rhizopods, are characterized by the looseness of their tissues, which allows the water to permeate every interstice, and to equalize the enormous pressure under which they live. When this pressure is removed, the fishes, with their flabby muscles, tender skins, and semi-cartilaginous skeletons, literally fall to pieces; they suffer from the decomposition and the dilatation of the air of the swimming bladder; the eyes are forced out of their sockets, and the scales fall off from the delicate skin. The mollusks present shapeless masses difficult of study. The crustacea seem to have been boiled, and their soft and thin shells resemble those of their shallow-water congeners just after moulting; many of the annelids and echinoderms look as if they had been digested by some of the larger deep-sea denizens, while the fragile types have lost their delicate appendages, or have become crushed in the ascent. Yet we know that a number of species of all these classes can thrive under differences of pressure due to such an extreme bathymetrical range as two thousand fathoms; but undoubtedly the individuals living at these enormous depths have found their way there very gradually, or ascend and descend from one level to another most leisurely, so as to become accustomed to differences in pressure.

Our information regarding the abyssal realms is far from complete, and our sketch of the natural history of the inhabitants of the floor of the ocean should be regarded only as a preliminary outline. Naturally, our knowledge of some of the groups is more extended than that of others, and the results obtained in any one case may differ radically from those reached by the study of less well known groups. As in the history of the fauna of any zoölogical province, our conclusions are constantly modified by the final results derived from a more careful study of some special case. There are of course certain rules applicable to all the inhabitants of the deeper regions, but they are few, and liable to constant modifications from our increasing knowledge.

In discussing the results of the "Blake" collections, I have availed myself most freely of the work done by other expeditions, as this is indeed essential for the proper understanding of the special facts examined. We are only on the threshold of our knowledge of the species and their exact distribution over the sea bottom; nevertheless the data of the various deep-sea expeditions seem to show that we know enough to form a general idea of the biological conditions under which these species exist, and that, judging from a few better known groups, our ideas are not likely to be materially modified by future researches.

This is especially the case with the West Indian fauna, and that of the east coast of the United States. We may safely assume that but little will hereafter be added to our notions of the association of the sponges, polyps, corals, echinoderms, crustacea, and mollusks, composing the West Indian deep-sea fauna, and making it in certain groups by far the richest in the world. The number of new forms from the West Indian region constitutes such a vast addition to our knowledge of the principal classes of invertebrates of that fauna as to revolutionize our ideas of geographical as well as of bathymetrical distribution.

No other region of the ocean bottom has yielded so abundant a harvest, and we have therefore no data elsewhere sufficiently complete for comparisons with regard to geographical distribution. But for ascertaining the bathymetrical distribution, and its bearing on the determination of the probable depth in which strata of former ages containing corals were deposited, the material at hand is of great importance.

I cannot give a better idea of the value of the collections brought together by the "Blake," than by contrasting the statistics of some of the groups before and after the Coast Survey explorations. I should state that the collections are as yet by no means fully worked out; but enough has been done, even in the groups least advanced, to show the wonderful richness of the collections, not only in new forms, but also in remarkable types of special interest.

Before the explorations of the "Blake" we knew nothing of the deep-sea fishes of the Caribbean Sea and of the Gulf of

Mexico. Less than fifty years ago there were not more than twenty known species of crustacea from the West Indian region. The "Blake" has added no less than forty new genera and 150 new species to those thus far described. Ten of the genera and nearly forty of the species belong to the well-known Brachyura, in spite of the fact that Stimpson and Milne-Edwards had, before the explorations made by the "Blake," apparently very fully worked out the species of this group from the dredgings of the "Hassler" and "Bibb"; sixteen genera and over sixty species belong to the less known Anomura; and there are fourteen genera and about fifty species of Macrura.

Among the mollusks the total number of littoral species recorded by Adams and D'Orbigny is 580, as compared with 461 collected by the "Blake." This number also includes 210 littoral species, while 251 are abyssal. The number of genera represented by the former is about 110, while some 98 genera are found in the "Blake" collection. These numbers are of course approximate.

The immense collections of echinoderms are peculiarly interesting. Of the deep-sea echinoderms the most striking are the Elasipoda, a new order of holothurians, established by Dr. Théel for the reception of these extraordinary and aberrant types, of which no less than fifty-two species were discovered by the "Challenger" expedition. Previous to that time three species of the group were known, one from the Kara Sea, and two subsequently found in the northern parts of the Atlantic by the Norwegian North Atlantic expedition. The "Blake" dredged about nine species of this remarkable order, three of which were unknown before.

There are now described eighty-three species of sea-urchins from the Caribbean fauna. Of these, eleven were added by the dredgings of Count Pourtalès in the "Bibb" and "Hassler," nineteen were discovered by the "Blake," and thirteen species previously known from other districts were obtained for the first time in the Caribbean and adjoining seas by the Coast Survey expeditions, so that the list of species has been more than doubled by the dredgings made since 1876.

The "Blake" dredged fifty-four species of starfishes, of

which forty-six were undescribed. As the total number of species does not exceed five hundred, the value of these additions to the group is readily estimated. Prior to the explorations of the "Blake," twenty-seven species had been described from the Caribbean region, so that the number of the species characteristic of the district has been nearly trebled; plainly showing that the deep-water starfish fauna is far richer and more varied than that of the littoral district.

The collection of ophiurans is perhaps the largest ever made. They seem to play a very important part in determining the facies of a fauna. They occur everywhere, at all depths, and often in countless numbers. I hardly think we made a single haul which did not contain an ophiuran. They often came up when the trawl brought nothing else. In some places the bottom must have been paved with them, just as the shallows are sometimes paved with starfishes and sea-urchins, and many species hitherto considered as extremely rare have been found to be really abundant. Most of the deep-sea Atlantic species obtained by the "Challenger" have been rediscovered in large numbers. Such rare species as *Sigsbeia murrhina, Ophiozona nivea, Hemieuryale pustulata,* and *Ophiocamax hystrix,* were found in plenty. As representatives of northern seas may be cited *Astronyx Loveni,* while the great rarities are represented by a single specimen of Ophiophyllum. Of *Astrocnida isidis,* of which only three specimens were known, we have half a dozen. A large Pectinura recalls the shallow fauna of the East Indies, while a new Ophiernus brings to mind the antarctic deep-sea forms. Finally, the supposed existence of simple armed Astrophytons is fully confirmed by the various species of Astroschema, and by a new species of Ophiocreas.

The diligent search of Pourtalès in the Straits of Florida, the "Hassler" expedition, the "Challenger" explorations, and the expeditions of the "Blake," have evidently brought up the majority of the species of ophiurans; for in the enormous mass of specimens gathered in the last "Blake" expedition and by the "Albatross" the number of new species was small.

It is noteworthy that the explorations of the "Blake" and the subsequent dredgings of the "Albatross" only added one species to the number of West Indian stalked crinoids. Three species of Pentacrinus were known before the explorations of the "Blake," — two of Rhizocrinus, and one of the strange Holopus. The importance of the collection of the free feather-star crinoids may be gathered from the fact that, while, according to Mr. Carpenter, the number of species of Caribbean Comatulæ is about fifty-five, three quarters of them were first obtained by the "Blake."

But although the species of stalked crinoids were known, the material formerly at the disposal of naturalists was most scanty, and some two dozen specimens of Pentacrinus represented probably the whole available supply. It was the fortune of the "Blake" to make the first extensive collections of this ancient genus; they were placed at the disposal of the late Sir Wyville Thomson, and finally passed into the hands of Dr. P. H. Carpenter, who worked out the anatomy of the genus in an admirable manner. In the Eastern Atlantic a very fine species of the genus (*P. Wyville-Thomsoni*) was discovered by Gwyn Jeffreys in the "Porcupine," off Portugal, in about 900 fathoms.

Innumerable fragments of stems of Pentacrinus, and portions of the arms, frequently came up in our earlier dredgings, but we were not fortunate enough until the last day of the first expedition to obtain a single entire specimen, though off Bahia Honda we dredged a young Holopus in excellent condition. When Sigsbee afterwards discovered, off Havana, the Pentacrinus ground, a short distance from the Morro Light, at a depth varying from 42 to 242 fathoms, he brought up about twenty perfect specimens of Pentacrinus of all sizes, besides a mass of fragments.

During the winter of 1879–80, Commander Bartlett also found Pentacrinus off Santiago de Cuba, and off Kingston, Jamaica, and a number of specimens of Rhizocrinus were obtained by the "Blake," but only a few were in perfect condition. Of Holopus a mutilated specimen was dredged. It was collected off Montserrat, and escaped my attention; as, being on the lookout for black Holopus, I did not notice this imperfect whitish specimen, which must have been alive, among the numerous Pentacrini with which it came up. During the second cruise our collection of Pentacrini became very extensive; we found them at Montserrat, St. Vincent, Grenada, Guadeloupe, and Barbados, in such numbers that on one occasion we brought up no less than one hundred and twenty-four at a single haul of the bar and tangles. We must indeed have swept over actual forests of Pentacrini, crowded together much as they may have lived, at certain localities, both in Europe and America, during the palæozoic period.

The monograph of Allman on the deep-sea hydroids of Florida gave us the first intimation of the wealth of forms which flourished in deep water, forming, as Allman says, a special province in the geographical distribution of the Hydroida. The collection was noted for the large number of undescribed species, and the small percentage which could be referred to forms existing on the European side of the Atlantic.

Previous to the deep-sea explorations we knew only the shallow-water reef corals. The expeditions of Pourtalès, of the "Hassler" and the "Blake," have revealed to us a whole fauna of simple corals separated from the reef district by a barren zone, with not a species in common between the two districts. There are now over sixty simple deep-sea corals known from the Caribbean district, — nearly as many species as there are from the reef area.

It is natural that, as we pass from the littoral to the continental, and finally to the abyssal regions, we should find a gradual diminution of those physical causes which we are accustomed to consider as influencing the variation of individuals, so that persistent types, as they have been called, may owe their origin either to an absence of modifying causes, or to an inherent tendency to retain unchanged their original organization. The animals we dredge from deep water cannot, from the nature of their surroundings, be affected, or only in a less degree, in the many ways which influence their shallow-water allies. We cannot suppose that they are subject at great depths to any of the causes which affect so powerfully the changing chromatophores of the littoral species; such adaptations as those which we find in the animals of the sargasso weed, for instance, or the littoral algæ, or those living on sandy or muddy or gravelly beaches, can hardly exist in the ooze of the abysses.

The habits of many of the deep-sea dwellers are still those of their shallow-water congeners, and yet their conditions of existence are so different that we can scarcely suppose them not to have equal importance. The mollusks, annelids, crustacea, and echinoderms which find shelter in the branches of the deep-water gorgonians, or the cavities of the abyssal Euplectellæ, cannot be subject to the attacks of so many enemies as those which live in shallower waters.

The metamorphoses of the deep-sea echinoderms, crustacea, annelids, and mollusks must to a great extent be adapted to their surroundings. Embryonic pelagic stages cannot be retained among the deep-water genera; these either pass through the so-called abbreviated metamorphosis within the egg, as in some crustacea and annelids, or after leaving the egg envelopes are kept in a kind of marsupium, as in some echinoderms; both these modes of development occur in the littoral and shallow-water species. Neither is it probable that the eggs of the deep-sea fishes are pelagic; they may be either too heavy to float, or in some families may be attached to the bottom.

Previous to the deep-sea explorations the collections made near the hundred-fathom line, or thereabout, were considered as belonging in "deep water," so that, when examining the early lists published by the English, Scandinavian, and American naturalists, we should bear in mind that they represent a fauna which scarcely extends beyond the limits of the littoral region as at present understood, and include only the few deep-water types which find their way to the junction of the littoral and continental regions. Of course the comparisons made with the strictly shore inhabitants, or those of adjacent bathymetrical belts, were often interesting, but had not the wide bearing of the results of later explorations.

The bathymetrical distribution of some of the more important types brings out strikingly the contrast between the faunæ of the submarine regions thus far recognized. An examination of the fishes obtained by the "Challenger," the "Blake," and the "Albatross," shows that twenty-six species have a vertical range of nine hundred fathoms or more. This vertical range is probably limited to the bottom, except, perhaps, in the case of pelagic fishes allied to deep-sea species, of which the habitat is always uncertain. The majority of fishes, to be sure, are bottom lovers when adult, but in larval stages, in the various phases grouped by ichthyologists in the family Leptocephalidæ, they are carried by the Gulf Stream and other currents, and spread far and wide over the ocean surface. Among the flat fishes a transparent pelagic embryo flounder known as Plagusia (see Fig. 78) passes under favorable circumstances into a deep-sea flounder; an allied species is known on the coast of Italy as Rhombodichthys.

It is an interesting problem to ascertain where the young of these fishes remain before they become permanent inhabitants of deep water. The same may be asked of some of the rarer pelagic fishes occasionally caught at sea, which undoubtedly are either fully grown deep-sea fishes or their young.

The greatest depth from which fishes have been dredged by the "Challenger" is 2,900 fathoms, and from that depth a single specimen (*Gonostoma microdon*) was brought up. The "Albatross" obtained from a depth of 2,949 fathoms a closely allied fish (*Cyclothone lusca*, Fig. 196), and four others. The "Talisman" secured one species from a depth of 4,255 metres, and the "Challenger" two from 2,750 fathoms, three from 2,500, and one from 2,650.

The larger part of the crustacea, both in the West Indian region and off the Atlantic coast of the United States, were brought from a depth of less than 500 fathoms. Out of about 100 species of Brachyura, only two were dredged below 500 fathoms; from about 75 species of Anomura, 22 were taken at or below 500 fathoms, five below 1,000 fathoms, and one below 2,000 fathoms; while among sixty species of Macrura thirty are recorded as taken below 500 fathoms, and thirteen below 1,000 fathoms.

The maximum range of the crustacea does not seem to be as great as that in other groups of invertebrates. In the Caribbean, only five species have a range of nearly 1,000 fathoms, and about the same number one of 500 fathoms.

The bathymetrical range of the mollusks is also connected with a wide geographical extension. According to Mr. Dall, if we consider the species dredged from the Atlantic Ocean north of a line drawn from Hatteras to Madeira, by all expeditions up to 1883, at greater depths than one thousand fathoms, we find that more than forty-two per cent live in some locality in less than one hundred fathoms.

These species of mollusks have apparently taken advantage of the uniform conditions of existence in deep water, and have extended their range far from their original littoral abode. There is a tolerable number of species, evidently unchanged, which occur all the way from a few fathoms, on the Florida coast, to two thousand fathoms in the adjacent deeps. A better knowledge of the littoral fauna of the tropics would undoubtedly increase this percentage. We also notice that the percentage of the genera or families peculiar to the continental and abyssal regions is small.

The sea-urchins and starfishes have their fullest development in the continental zone, and there we find already many of the genera and families which have given so characteristic an aspect to the fauna of deep waters. Beyond that region live the eminently deep-sea types of the Pourtalesiæ and Ananchytidæ, associated with a few starfishes and the strange order of holothurians, the Elasipoda. The ophiurans appear, of all the echinoderms, to flourish best in the deepest waters from which members of the class have as yet been dredged. The bathymetrical range of many of the sea-urchins and ophiurans is very great, and extremes of depth extending to two thousand fathoms or more are not uncommon.

The stalked crinoids, as has been shown by Carpenter, are not strictly abyssal types; on the contrary, seventy-five per cent of them have been brought up from depths of less than five hundred fathoms, — somewhat deeper than the limit of the continental zone. As stated by Carpenter, out of the thirty-two recent species of stalked crinoids, nine species may be called littoral, living as they do at depths of less than one hundred fathoms.

Comatulæ were dredged at fifty-seven out of the two hundred

stations occupied during one season's work. Nearly all of them were in comparatively shallow water, *i. e.* in depths of less than two hundred fathoms. On three occasions the depth exceeded three hundred fathoms.

These facts agree well with the results of the "Challenger" dredgings, which yielded Comatulæ at twenty stations only where the depth was more than two hundred fathoms. One may fairly conclude, therefore, that these animals are essentially inhabitants of shallow water. The crinoids form a striking exception to the rule, which holds good among many of the other groups, that the more ancient types also have a wide range in depth.

The bathymetrical distribution of the corals is such that we can readily separate the species found in depths of less than one hundred fathoms, where they live in the region of *débris* which lies between the reefs and the rocky or muddy bottoms. But here again there is no sharp line of demarcation in the distribution between the continental and the deeper zones, though the abyssal regions contain a comparatively smaller number of species than the continental slope. They flourish upon the continental slope only on sea bottoms which are free from accumulating silt, and remote from flat muddy shores and from the influence of great rivers; the branching types prefer a rocky or stony bottom, while the simple types thrive on shelly or oozy bottom. It is on this slope that we also meet with the greatest number of novelties among the gorgonians and pennatulids, while specially characteristic of the deeper regions is the family of Umbellulæ.

The calcareous and horny sponges, of which our commercial sponge is a good representative, are eminently littoral forms. Beyond that depth the bright-colored sponges are replaced by the hosts of siliceous sponges which live buried in the mud, some of them anchored by their bundles of gigantic spicules deep in the ooze, which also envelops them in a thick coating of fine mud so closely held by the network of the skeleton that careful preparation alone brings out the wonderful beauty of their structure. An Euplectella when first brought up looks like a mere mud-lined cylinder, and gives no idea of the exquisite tracery formed by the siliceous skeleton.

The sponges also seem specially to dwell upon the continental slopes, and here it is that the kingdom of brightly colored sponges displays its splendor of yellow, orange, red, and brown. The sponge zone is comparatively narrow on the bank of Florida, where perhaps it takes its greatest development in the districts explored by the "Blake;" it disappears at about one hundred and fifty fathoms, sometimes before, particularly where the bottom affords favorable conditions for the deposition of silt or ooze, which is destructive to the development of all except the siliceous sponges.

The Lithistidæ and Hexactinellidæ do not occur in the littoral zone, while the other families, though often extending into deep water, also run into the littoral zone, but take their principal development between one hundred and two hundred fathoms.

The dredgings of the "Blake" reached from shallow water, generally within the hundred-fathom line, to the abyssal depths of the same area. These dredgings therefore give us terms of comparison for the inhabitants of all depths of the same region, many of which are missing from the collections of the other deep-sea explorations, as they ceased work when approaching the shore line.

We are thus able to trace far more accurately than we could from other collections, not only the species which are merely littoral and have migrated into deeper water, often at a considerable distance from their original littoral habitat, but also those which after migration have become modified so as to form the characteristic faunal inhabitants of the continental and abyssal regions, and those cosmopolitan species, assumed to be of arctic or antarctic origin, which have an immense geographical range over the whole bottom of the Atlantic and Pacific oceans. The last may be considered stragglers or colonies, which have found their way, towards both the littoral and abyssal regions, into faunal districts not strictly their own, according to the distance of deep water from the shores, or the nature and direction of currents. We may thus get a most striking contrast between the faunæ of adjoining littoral, continental, and abyssal regions. This is shown by palæontological evidence from districts corresponding to the shallower continental regions of our day.

In the experience of the "Blake" the greatest wealth of specimens, or the principal treasures of the expedition, were not dredged from the deepest waters of West Indian or Atlantic areas. It was mainly upon the continental slopes, near the five-hundred-fathom line, where food is most abundant, or the slopes are washed by favorable currents, that the richest harvests came up in the trawl. Several places really phenomenal from their richness were met with by the "Blake," — off Havana, to the westward of St. Vincent, off Frederichstæd, off the Tortugas where the Gulf Stream strikes the southern extremity of the Florida Reef, and off Cape Hatteras. We might also name the remarkable spots found by the "Challenger" off Japan and off Zamboanga, and the rich dredgings of Pourtalès on the plateau which bears his name. We may safely say that the abundance of life in the many favored localities of the ocean far surpasses that of the richest terrestrial faunal districts. The most thickly populated tropical jungle does not compare in wealth of animal or vegetable life with a marine district such as a coral reef, or some of the assemblages mentioned above.

It will be impossible to give a good picture of the animals which make up the fauna characteristic of certain well-defined regions until we have the completion of the reports by the different specialists who have kindly consented to work up the collections of the "Blake." We may, however, call attention in a general way to their geographical and bathymetrical distribution. There can be no greater difference, for instance, than that which exists between the animals associated in deep water on the rocky bottom upon the southern slope of the Florida Reef, on the Pourtalès Plateau, with its predominance of corals, Rhizocrinus, and starfishes, and those found in the calcareous ooze of the trough of the Gulf Stream (lamellibranchiates, holothurians, etc.); and again in the association of the masses of Gorgoniæ, Saleniæ, and Terebratulæ, off the north coast of Cuba, brought up in a single haul of the trawl. Nor can there be a greater contrast than between the inhabitants of the pteropod ooze in deep water off the west end of Santa Cruz, with its preponderance of Phormosomæ, of Asthenosomæ, and Hyalonemæ, and those of the forests of Pentacrini and Gorgoniæ, and the accompanying Comatulæ and Ophiuridæ, living in such numbers on the windward coast of St. Vincent.

We may contrast, again, the deep-water fauna off the Tortugas, in the coral ooze, mainly made up of a most characteristic association of fishes and crustacea, with the hauls in deep water at special localities, consisting entirely of thousands of specimens of single species, either of ophiurans, or of sea-urchins, or of feather-stars, or of crustaceans, or of gorgonians.

Take again the bottom around the ridges between the West India Islands, or that along the course of the Gulf Stream off the Carolinas, which are swept nearly clear of all animal life, and compare their inhabitants with the rich and varied fauna of the same depths upon the continental shelf farther north, and along the western shelf of the Windward Islands, on the lee side, in the Caribbean; or compare these faunæ in turn with the mass of animal life, mainly composed of gorgonians and calcareous and horny sponges, found upon the broad plateau on the west of Florida and on the Yucatan Bank; there can be no greater contrasts than those of the narrowly circumscribed areas I have mentioned, where all the animals belong to the West Indian fauna taken as a whole. This clearly indicates radical faunal contrasts in very limited areas, which differ principally in the character of the bottom, and where the physical conditions, such as temperature, depending mainly upon currents and winds, are in striking opposition within comparatively moderate distances.

But by far the most marked contrast is perhaps presented by the reef fauna to that which immediately follows it towards deeper water. None of the corals of the most abundant families or species characteristic of the West Indian reefs extend to any considerable depth, and simple corals, which form so large a portion of the deep-sea fauna, are not represented at all in the Florida reef fauna. It was on the slopes of the rocky plateau stretching into deep water off the Florida reefs that Pourtalès first dredged the extraordinary assemblage of ancient animals which constitute the continental fauna, succeeding in depth the reef fauna just mentioned. The contrast between the littoral fauna of the tropics and that of the continental and abyssal regions is far greater than that between the inhabitants of the same regions in the temperate or arctic provinces. This is readily explained by the circumstance that the cold water of the abyssal regions, with its characteristic animals, approaches nearer the shore as we go north within the continental region, so that the littoral fauna of the arctic circle lies practically under the same conditions of temperature as the abyssal in the tropics, or the continental in the temperate zones. That is, the divisions of these faunal regions are to be determined more by temperature than by depth, although of course the temperature depends upon the depth and upon the currents of the ocean. Below a depth of seven to eight hundred fathoms, corresponding to a temperature of 40° F., we pass into the abyssal regions, while upon the continental slope at a depth of about 150 fathoms we reach the lower limit of the littoral region.

One of the first points noted by Lovén in reference to the few deep-sea types occasionally brought up from various quarters of the Atlantic was their wide geographical range; and he first distinctly formulated the theory of the uniformity of an abyssal fauna extending in the Atlantic from the arctic to the antarctic regions, with a somewhat modified fauna at the two poles, — a theory which has been slightly changed by later deep-sea explorations. Lovén's theory seemed to give a most natural explanation of the marked similarity, often noticed by various naturalists, between a number of the -arctic and antarctic invertebrates. It was therefore of the greatest interest when Pourtalès dredged in the deep water of the Straits of Florida the little Rhizocrinus discovered by Sars on the coast of Norway, and when subsequent explorations of the "Blake" brought to light a large number of boreal types in the deep water of the Caribbean district, and off our eastern coast. Professor Smitt, who examined our collection of the Bryozoa from the West Indian district, speaks of the interest he felt in finding well-known Scandinavian forms among these tropical and antarctic types. The range of many of the Bryozoa is very wide. More than ten Caribbean species are found in the North Atlantic, and an equal number extend to the arctic regions; eight are Australian, and four belong also to the Red Sea.

About as many species are identical with those of the Antarctic Sea and the southern extremity of South America. The species which attain the greatest depth are usually those which have a very wide geographical distribution, generally with an arctic or antarctic connection, or they may be species dating back to the tertiary and cretaceous periods.

The similarity of the holothurians of the arctic and antarctic regions has been recognized by Théel, but no species are common to the two seas; it is therefore not probable that there is any interchange between the fauna of those distant regions, although in former ages such a connection may have existed from the wider geographical range of their progenitors; it is interesting to note in this respect, that in the Psolidæ, which find their way into very deep water, and have representatives in the tropic, temperate, and arctic zones, it is often most difficult to draw the specific limits. Still there are slight differences, indications of the changed physical conditions and various modes of life, which have caused the species to disappear from the intermediate localities. The same resemblance is noticed among the sea-urchins, the starfishes, and the ophiurans.

One of the most remarkable instances of the geographical extension of some genera is that of certain species of the family Lithodina. Professor Sidney I. Smith says: "These crustacea have been known as inhabitants only of the arctic and antarctic regions, living in the littoral zone; but now they have been found under the tropics, the only difference being that in this latter locality they have contrived to find congenial conditions of existence by abandoning their shallow-water life and betaking themselves to the cool depths of over 1,000 metres. This fact is not without its interest, showing us how some forms can spread from the frozen seas of the north to the seas of the tropics, and so from one pole to the other; altering their conditions of life as necessity demands, and resuming their old habits when the opportunity to do so again occurs."

Several species of sea-urchins are cosmopolitan; a number thus far seem peculiar to the Atlantic or to the Pacific, and these types all have a great bathymetrical distribution, or are representatives of fossil families that go back to the palæozoic, secondary, or tertiary times. This extension of geographical range in the case of so many of the species of the Caribbean fauna is most instructive. As has been observed in several groups of invertebrates, and in fishes, the presence of identical species on the two sides of the Isthmus of Panama points to a comparatively recent communication between the Atlantic and Pacific, while the presence of cosmopolitan species at such distant points as the Caribbean, Australia, and the Red Sea indicates a connection which could have been effected only by migration on the floor of the ocean or in the track of currents.

The sponges apparently have a wide geographical distribution, many of them being cosmopolitan. A number of mollusks also have an extraordinary geographical range, from Northern Europe to the Cape of Good Hope or to Patagonia. Others are found in the seas of Great Britain, at the Cape of Good Hope, and in the Southern Ocean. Others again are denizens of the arctic and antarctic seas, or extend from the northern parts of the Pacific to the Kerguelen Islands.

A number of species of deep-sea corals and gorgonians extend northward in deep water from the Caribbean district along the east coast of the United States. A few species of simple corals like Flabellum and Fungia have a great geographical and bathymetrical range. Half a dozen species of corals are common to the northern seas of Europe and the Straits of Florida. From the geographical distribution of the corals, and their affinity with the tertiary fossils of Italy, Pourtalès came to the conclusion that the tertiary deep-sea fauna of Europe has as it were migrated westward and maintained itself, while the greater part of the contemporaneous forms of the West Indian deep sea have become extinct.

The collections obtained by the "Blake" in the Caribbean district are superior, as regards the number of duplicates, to those made by the "Challenger." Many species occur, not only in large numbers, but also at several localities; so that it has been possible to study their range of variation in a more satisfactory manner than hitherto. This opportunity has proved of immense value in revealing the existence of many intermediate forms between types which were considered quite distinct.

Many groups are remarkable for the variety of their forms, so that it is almost impossible to apply to them any classification, even that regarded as best established. From the study of these groups, most interesting morphological and palæontological results have been derived. Some of these are discussed in connection with the account of the different zoölogical groups. As the corals of the West Indies have been carefully studied by Pourtalès, we may dwell more at length on the relations of that fauna to their precursors in the tertiary period.

The corals of the European tertiaries are so well known from the works of Milne-Edwards, Haime, Reuss, Seguenza, Duncan, and others, that we can compare the living West Indian coral faunæ, both littoral and abyssal, with that of the European tertiaries. The resemblance is a striking one, and we may safely, from analogy, reconstruct the physical conditions which existed in the European tertiary seas, and picture to ourselves the depth of the water, the purity of the sea, and the intense aeration of the waters, far from great bodies of fresh water, which must have prevailed in those days over areas where either coral reefs or a deep-water fauna flourished.[1]

Fewer deep-sea genera are common to the tertiary and living faunæ of the West Indies than to the European tertiary and the living West Indian fauna. This may be due to smaller changes of level in the latter region than in Europe. Yet if we take into account the fact that the numerous West Indian extinct genera belong to families of deep-sea corals, we may safely conclude that there have really been important changes of level in the West Indian area. The presence of European cretaceous fossils

[1] The similarity in the deep-water types and their fossil representatives may not invariably mean existence under identical conditions. We have the most satisfactory evidence that the crinoids of the silurian deposits of the State of New York flourished in shoal-like areas, and that during the jurassic period their occurrence on the coral reefs of that time showed these ancient crinoids to have lived in much shallower waters than their recent allies, the Pentacrinus and Rhizocrinus of the West Indies. The occurrence of the recent stalked crinoids in such deep water as compared with that of the palæozoic period may be interpreted to represent the conditions necessary for the maintenance of the type down to the present day. In the present epoch depth represents, as has been suggested by Pourtalès, the great pressure to which the heavy atmospheres of earlier periods subjected the animals of those days, and thus perpetuates conditions recalling those of the shoal waters of early ages.

in the West Indian miocene is not more anomalous than is the occurrence in the deep water of the West Indian seas of living species which perhaps characterized the Sicilian tertiaries. The beds, forming raised terraces such as those of the Barbados and of other islands of the Caribbean, though they seem to be the direct continuation of the coral beds now growing, yet also give us the measure of the physical changes which must have taken place in the West Indian regions about the end of the cretaceous, at the time of the separation of the Pacific Ocean and the Caribbean Sea.

The absence of single simple species of corals in the Caribbean district within the reef area distinguishes this fauna at once from that of the reef regions of the Pacific and Indian oceans, in which are found in shoal or moderately shoal water several species of simple corals, like Flabellum, many Fungidæ, and others, besides genera and families not represented in the West Indies. Yet the bathymetrical distribution of the West Indian species gives us an approximate idea of the depths at which some of the fossiliferous strata of the cretaceous and tertiaries containing corals were probably deposited.

Pourtalès, who thoroughly studied the deep-sea corals of Florida, was of the opinion that some of the miocene, pliocene, and pleistocene strata of Messina, of which the fossils have been so carefully described by Seguenza, were deposited in a depth averaging 450 fathoms, and ranging from about 200 to 700 fathoms. In the neighborhood of Vienna we may trace from Reuss's monographs the fluctuations of depth which have taken place between the deposition of the different strata. The miocene beds, in which there are numerous astræans associated with Porites, are shoal-water deposits; while the strata containing Turbinolidæ, Oculinidæ, and Eupsammidæ were formed in deep water.

The West Indian tertiary corals are not sufficiently known to permit us to reconstruct from them alone the past history of the ancient Caribbean seas. Duncan observed that, on some islands, such as Antigua and Trinidad, only reef species flourished. This shows conclusively that in other places there must be deep-sea deposits of the tertiary period which have not yet been brought to light. It is possible that the massive types of the West Indian miocene, such as the Asterosmiliæ and others which have no analogues at the present time, may have been living in the shoal water protected by reefs in the same way as the Fungiæ of the Pacific, or some of the unattached compound corals, as Manicina or Isophyllia of our coral reefs.

According to Mr. Dall, a large proportion of the miocene and even pliocene fossils of this country and of Sicily still exist in a living condition near our shores. They are found principally in the continental region. There are not, however, a sufficient number of antique types to characterize the deep-sea molluscan fauna as archaic, and none of them are as remarkable as the Australian Trigonia, the Caribbean Pleurotomaria, or the Indian Nautilus.

XV.

SKETCHES OF THE CHARACTERISTIC DEEP-SEA TYPES. — FISHES.[1]

THE collections of the earlier deep-sea expeditions consisted almost exclusively of invertebrate animals, and it was not until the publication of the "Challenger" results that any large number of deep-sea fishes became known. The first extensive contribution to our knowledge of the vertebrate inhabitants of the great depths of the sea was made by Dr. Günther of the British Museum, in 1878. He printed in the "Annals and Magazine of Natural History" a series of papers containing descriptions of some species of fishes which had been obtained by the "Challenger."

The deep-sea fishes, as a whole, although distinguished by marked peculiarities, consist of types not wholly unfamiliar to the ichthyologist. Many of the characteristic abyssal families have representatives in the inshore faunæ, less strongly specialized perhaps than their allies in the abysses, but still structurally the same. Others had in former years become known, from dead individuals which floated to the surface or drifted ashore. The latter have usually been designated as "pelagic forms," and until the existence of a deep-sea fauna was revealed, the problem of their origin was much less intelligible than it is now.

Even now, the distinctions between the inhabitants of deep water, those of the middle depths, and those of the surface strata of mid-ocean, are not strongly defined. Such are the imperfections in the methods of trawling and dredging, that the naturalist, when he has sorted out the fishes from his nets after

[1] I am indebted to Professor Goode and Dr. Bean for notes upon the Fishes. The figures are taken from a Memoir preparing on the Deep-Sea Fishes of the East Coast of the United States by Goode and Bean, based upon the collections of the "Blake" and of the U. S. Fish Commission.

a haul in mid-ocean, is entirely at a loss to know where his captures have been made. If he has taken a flounder from a haul in 800 fathoms, or finds a macruroid, a brotuloid, a berycoid, a synodontoid, or a nemichthyoid in a net which has been below

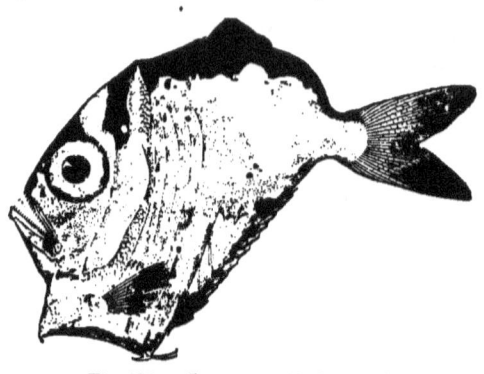

Fig. 195. — Sternoptyx diaphana. ⁴⁄₅.

the two-thousand-fathom line, he feels tolerably sure that he has brought it up from the bottom. But who shall say where those which like Argyropelecus, Sternoptyx (Fig. 195), or Cyclothone (Fig. 196), having allies among the pelagic fishes in the same net, have come from ? They may have come from the bottom, or they may have become entangled in the meshes of the trawl when but a few fathoms below the surface, in its ascent or descent. Many of the deep-sea fishes undoubtedly lead a most active life in spite of their cartilaginous bones and feeble muscular system, being kept efficient perhaps by the enormous pressure under

Fig. 196. — Cyclothone lusca. ⁴⁄₁. (U. S. F. C.)

which they live. The absolute calm of the abyssal regions may be the cause of the extraordinary development of some of the tactile or other organs of sense occurring in different parts of the skin, usually on the head or upon the lateral lines; some of these may be, as has been suggested by Leydig, accessory eyes, or phosphorescent organs. The accessory eyes may perform the part of bull's-eyes, thus constituting, according to Dr. Günther, "a very deadly trap for prey, one moment shining that it might attract the curiosity of some simple fish ; then extinguished, the simple fish would fall an easy prey." Some of the long filamentous organs are phosphorescent, while others are merely tactile.

Many surface fishes also descend to considerable depths. In fact, the migration of our coast fishes is one of the most important problems which the fisherman has to solve, and one of which we as yet know but little. There seems to be no serious obstacle to extensive bathymetric movements on the part of fishes. The silver hake, which is abundant all summer long at the surface on the New England coast, has been taken from 487 fathoms, and appears to live in September and October at considerable depths off Southern New England. There is reason to believe that the mackerel, menhaden, and the bluefish also go down below the hundred-fathom line in winter.

The fishes of the abyssal realm are very distinct from those of the surface faunæ. It is safe to say that there are more genera common to the seas of Australia and North America than to the littoral and abyssal faunæ off the Atlantic coast of the United States, — excluding the pelagic types,· many of which are cosmopolitan. Indeed, of the sixty or more genera which have been dredged below 1,000 fathoms in any sea, only one has been found in less than 200 fathoms on our own coast, and four within the two hundred-fathom line in any sea, even in polar regions. Of the same assemblage, only seven occur anywhere in less than 300 fathoms, and down to 500 fourteen are added to the list. These fourteen genera represent ten families. Out of the thirty-four family groups which are represented below 1,000 fathoms, or in mid-ocean beyond soundings, only five are represented in any in-shore fauna, even in circumpolar regions.

We have now considered the composition of the abyssal fauna, as found at the greatest depths. A glance at its upper limits may also prove instructive; we find below the hundred-fathom line, and within the limit of 500 fathoms, a very heterogeneous assemblage. Well-known surface species inhabit at times water of considerable depths. The cod goes below 100 fathoms; the halibut and the Newfoundland turbot go below 300, and the haddock apparently to 500, on the New England coast. Hake are also deep-sea lovers, being recorded at a depth of over 304 fathoms. One of the species of Phycis (*P. regius*) from 233 fathoms was discovered to be electric, giving quite a strong shock to Commander Bartlett and me. The goose-fish and the hag go down at least over 350, the " Norway haddock" to more than 150 fathoms. The swordfish, when attacked at the surface, is able to "sound" with ease and rapidity to a depth of 500 or 1,000 feet, arriving at the bottom with such force as to imbed its sword at full length in the mud, and there seems to be nothing to prevent powerful swimmers from visiting the bottom at any time when the conditions of temperature will permit. Scopelus, one of the most common pelagic fishes, may live at considerable depths: it comes up to the surface mainly during calm nights.

The number of representatives of shallow - water families dredged below 100 fathoms and down to a depth of 500 fathoms is quite large, but diminishes rapidly below that depth, two or three extending only to 700 fathoms, and an equal number to 1,000 and 2,000 fathoms.

To the bottom-living species which may have made their way gradually down to deep water upon the continental slopes belong preëminently the flat fishes. Fourteen species have been detected on our Atlantic coast, living beyond the hundred-fathom

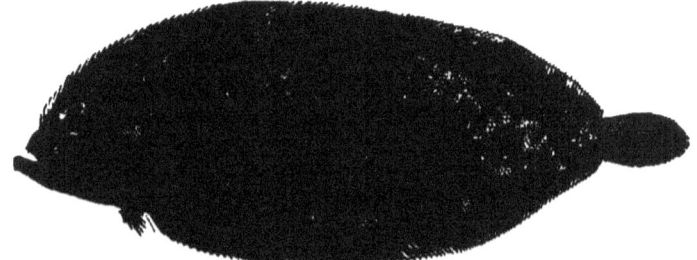

Fig. 197. — Monolene atrimana. About ½.

line. One of them (Monolene) (Fig. 197) comes from 300 fathoms, and three genera occur well down toward the thousand-fathom line. The pole flounder ranges beyond this limit, and breeds in deep water. It has the cavernous skeleton of the deep-sea fishes. In Bedford Basin, Nova Scotia, and in adjacent waters, it lives at depths of about 15 to 20 fathoms, and yet individuals captured there exhibit the peculiarities of abyssal types.

The flat fishes are represented by at least two genera fossil in the schists of Glaris, believed to have been the bottom of a deep sea, and in the clays of Sheppey are found fossil the three genera Gadus, Merlucius, and Phycis, — types which rarely go below 1,000 fathoms. Of the eleven recognized families of anacanthian fishes (flat fishes, cods, and the like), all save four are known from the abyssal fauna. The brotulid forms allied to the cods represent a dominant abyssal group.

Among them may be mentioned Barathronus (Fig. 198)

Fig. 198. — Barathronus bicolor. About ⅔.

(1769 fathoms), a small-eyed fish with marked colored bands upon its flanks, and Barathrodemus (Fig. 199) (647–1395 fathoms), a

Fig. 199. — Barathrodemus manatinus. About ⅔.

cusk-like fish. One of the most interesting forms of the Brotulidæ is Aphyonus, with rudimentary eyes, one species of which,

Fig. 200. — Aphyonus mollis. About ⅘.

having no visible eyes, was obtained by the "Challenger" at a depth of 1,400 fathoms, south of New Guinea; another, *A. mollis* (Fig. 200), by the "Blake," in 955 fathoms. This fish is covered by a flaccid, scaleless skin, is toothless, and has its head covered with a system of wide muciferous canals, the dermal bones being almost membranaceous. It is either a very ancient or a very degenerate type, but bears a remarkable superficial resemblance to its ally, Lucifuga, which inhabits the subterranean waters of caves in Cuba, and has lost the use of its eyes.

The typical family of cods (Gadidæ) is also numerously represented in the depths of the sea; those forms which descend to the greatest depths being usually of a more elongate form than the brotulids, and with a small, often filamentous, first dorsal fin.

The Ophidiidæ (*Ophidium cervinum*) (Fig. 201) are elongated Gadoids.

Fig. 201. — Ophidium cervinum. About ½. (U. S. F. C.)

The Lycodidæ are abundant in the polar waters and lesser abysses of the North Atlantic and Pacific, and occur also where the Atlantic abysses merge into the Antarctic.

The macruroids (Fig. 202) are characteristic abyssal forms, and both specifically and individually are exceedingly numerous at all depths below the hundred-fathom line. Seventy-five per cent at least of the fishes brought up in the trawl from the abyssal regions are members of this family. Macrurus is rare below 1,000 fathoms, only one species, *M. Bairdii*, having straggled below this limit. It is more abundant inside the five hundred-fathom line, and Steindachneria, a macruroid with a high differentiated first anal fin, has been obtained by the "Albatross" in 68 fathoms. The species and individuals of Coryphænoides and Bathygadus (Fig. 203) are as numerous below 500 fathoms as those of Macrurus are above it. The cavernous structure and membranous texture of their skeletons are very marked, and they seem, through their elongate forms, tapering tails, immense heads, and strongly armed bodies, to be especially adapted

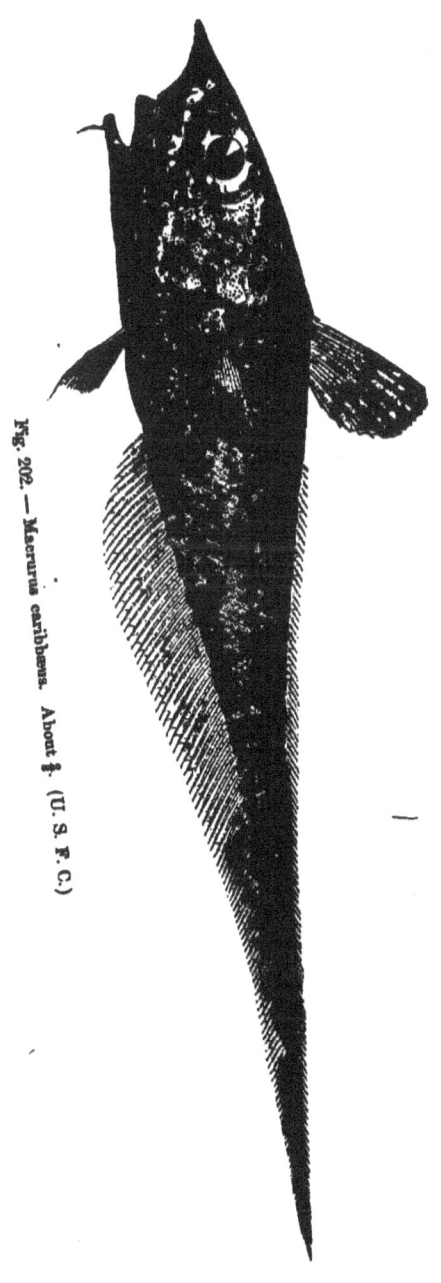

Fig. 202.—Macrurus caribbaeus. About ⅔. (U. S. F. C.)

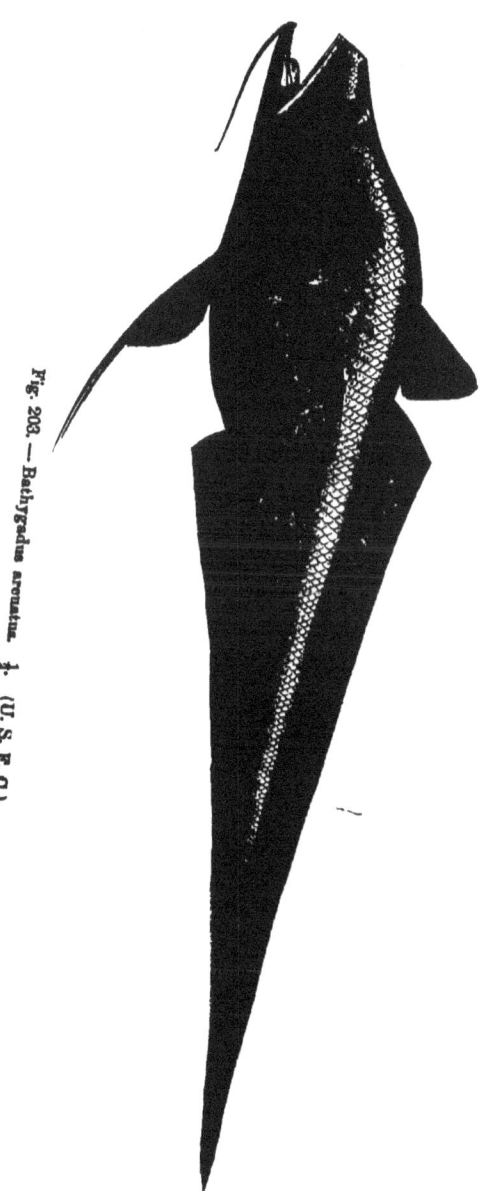

Fig. 203.—Bathygadus arcuatus. ⅓. (U. S. F. C.)

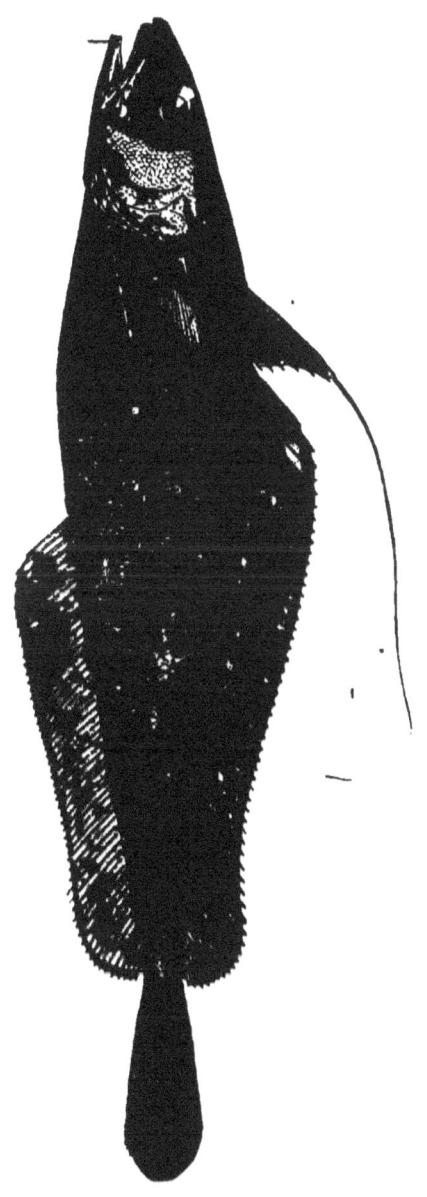

Fig. 204.—Phycis Chesteri. ⅔. (U. S. F. C.)

to life in the ooze and slime of the bottom. *Macrurus Bairdii* and *Phycis Chesteri* (Fig. 204) are the two most common fishes of the continental slope, where they occur in immense numbers, and breed at depths varying from 140 to 500 fathoms.

The family Bregmacerotidæ, hitherto known only through a single species, a native of the Indian Ocean, appears adapted to living at considerable depths. The discovery by the " Blake " of a species (the long-finned *Bregmaceros atlanticus*) (Fig. 205)

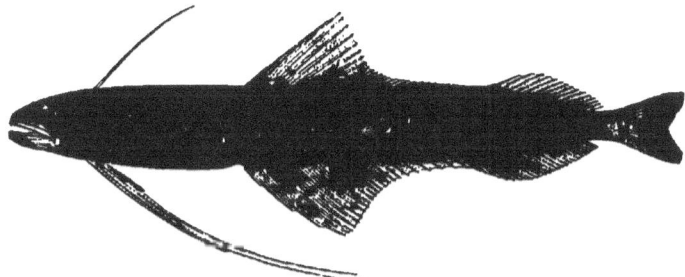

Fig. 205. — Bregmaceros atlanticus. ⅔.

of this old-world genus in the Gulf of Mexico, at a depth of 305–390 fathoms, is very interesting to ichthyologists.

Certain groups of the blennies, gobies and the like, often send stragglers down to the lesser abyssal depths. They are forms with more or less elongate bodies, and low, feeble vertical fins, adapted neither to free swimming nor to the pursuit of prey at the surface. They are, in fact, bottom feeders, somewhat sluggish in habit, and usually live among stones and hide in crevices; while, as a rule, fishes like the perch, the sea-bream, and the mackerel, belonging to groups with compact, short bodies, powerful fins, and boldly predaceous disposition, do not descend to great depths, and do not wander far from the coast waters. The Berycoidea, the first group of bony fishes to appear upon the geological horizon, occurring early in the cretaceous, are represented in the deepest dredgings of the " Albatross " (2,949 fathoms) by a species of Plectromus. (Fig. 206.) The Norwegian deep-sea expedition found a species of Beryx, and *Beryx splendens*, a magnificent brilliant scarlet species, known hitherto only from Madeira, was one of the most important captures of the " Albatross," in 460 fathoms.

Fig. 200. — Plectromus suborbitalis. ¼. (U. S. F. C.)

The snappers and groupers of the tropics surely range below one hundred fathoms, but it seems hardly appropriate to regard any of the true percoids, or any of their very near allies, as really abyssal in habit.

Some of the scombroids seem to inhabit deep water, especially the Trichiuridæ, the so-called cutlass-fishes, which may be considered a deep-sea group. They are long, compressed, of glistening silver color; they date back to the chalk of Lewes and Maestricht, and occur in the eocene schists of Glaris. A number of pelagic scombroids have been taken under such circumstances as to render it probable that they descend to considerable depths. The lumpsuckers (Liparidæ) are well represented by four genera, which have undergone extreme modifications characteristic of abyssal forms. They have soft, cavernous skeletons, immensely developed mucous canals, and are soft and flaccid in the extreme. The family of lump-fishes (Cyclopteridæ) is represented below the hundred-fathom line off the Atlantic coast.

The "ribbon-fishes" may be named with the abyssal groups, although they have never been dredged at any considerable depth, but are known solely from individuals stranded upon the shores or found at the top of the water. The largest of the ribbon-fishes is capable of rapid motion at the surface, and is probably the animal which has most often been taken for the sea-serpent. The "Bermuda sea-serpent," *Regalecus Jonesii*, was seventeen feet long, and swam with great velocity through the surf, and dashed itself upon the shore. It seems altogether reasonable to believe that these fishes live at comparatively moderate depths, like the members of the family Trichiuridæ.

Among the bottom-loving groups, the sculpin descends to 732 fathoms; its representatives go back to the tertiary formations.

The scorpænoids descend to 440 fathoms. Scorpæna occurs in the eocene of Oran.

The blennies are still represented at a depth of 471 fathoms.

The gobies have a representative in deep water, Callionymus (Fig. 207), a huge sea-robin-like fish. The discovery of a mem-

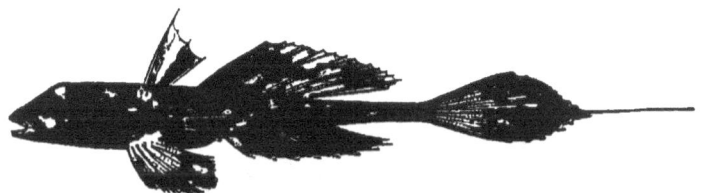

Fig. 207. — Callionymus Agassizii. About ½.

ber of this old-world family in the Gulf of Mexico, at a depth of 340 fathoms, is one of the noteworthy features of the "Blake" exploration.

We should also mention the tile-fish dredged off our Middle Atlantic coast in deep water, the remarkable *Lopholatilus chamæleonticeps*.

Chiasmodon niger (Fig. 208) is a species which has been

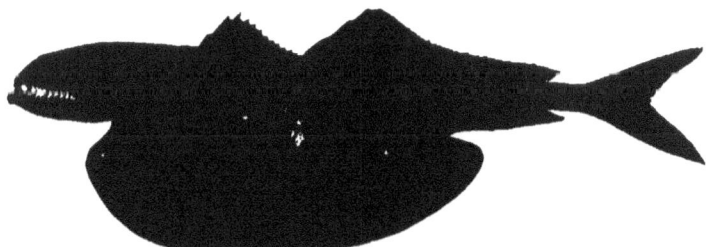

Fig. 208. — Chiasmodon niger. About ½. (U. S. F. C.)

often described, but its common name, "the great swallower," is so characteristic that we may here recall it to memory. It is able to take in fishes fully half as large as itself. Günther places it in 1,500 fathoms. Most of the specimens known have been collected at the surface, and there seems to be a reasonable probability that this genus inhabits intermediate depths, since mid-depth fishes only have been found in its stomach.

The gurnards have also representatives in deep water, if the remarkable new genus Hypsicometes is one of its members. This has been obtained both by the "Blake" and by the "Albatross" at various depths from 68 to 324 fathoms, and four species of the family touch the hundred-fathom line or go below it.

The Agonidæ are represented in 324 fathoms by one species of Peristedium (Fig. 209), remarkable for its branching barbels,

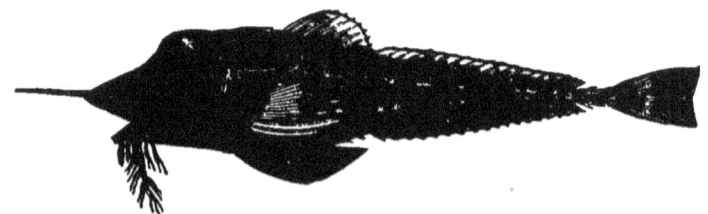

Fig. 209. — Peristedium longispatha. About ⅔.

and three others found between 140 and 300 fathoms, — all the result of recent American explorations.

It is worthy of note, that the characteristic abyssal families are apparently offshoots of free-swimming species of active habits, which have, in the course of time, become gradually acclimated in the depths of the sea. Their approach to great depths would appear to have been in vertical lines, rather than upon the slopes of the ocean bottom.

One of the most aberrant types, Notacanthus, was obtained by the "Challenger" from a depth of 1,875 fathoms. *N. phasganorus* was taken from the stomach of a shark killed on the Grand Bank of Newfoundland.

Many members of the group of Pediculati are often met with swimming on the surface. They are species whose habits seem to have become modified to those of deep-sea fishes, while they apparently retain the characteristics of their surface allies, the most familiar representatives of which are the goose-fish (Lophius) and its allies (Malthe and Pterophryne). *Lophius piscatorius*, the common goose-fish of the North Atlantic, descends to 365

Fig. 210. — Nest of Pterophryne. About ½.

fathoms. Pterophryne, "the marbled angler" of the Sargasso Sea, is specially adapted to live among the floating algæ, to

Fig. 211. — Antennarius. ⅔.

which it clings with its pediculated fins, and in which it intertwines its gelatinous clusters of eggs. (Fig. 210.) Its ally,

Antennarius (Fig. 211), has become adapted to life on the bottom, and is found nearly down to the hundred-fathom line. *Chaunax pictus*, a closely related genus, was taken by the "Blake" in 288 fathoms. The Ceratiidæ are the only pediculates which are exclusively and characteristically abyssal. Melanocetus, a deep-sea Lophius in appearance, ranges from 360 to 1,850 fathoms; the "Blake" took it in 992 fathoms.

The Alepocephalidæ, the Halosauridæ (Fig. 213), and Chauliontidæ (Fig. 214), are families which have become perma-

Fig. 215. — Ipnops Murrayi. About ½.

nent residents on the bottom. To the former belongs *Alepocephalus Agassizii* (Fig. 212), a magnificent fish which attains a length of at least three feet, is covered with silvery scales, and is noted for its large eyes; while allied to the scopelids, but inhabitants of deep water, belong certain genera, as Ipnops (Fig.

Fig. 217. — Bathypterois quadrifilis. About ⅔.

215), Bathysaurus (Fig. 216), with its huge dorsal fin and fine teeth set in many rows, Bathypterois (Fig. 217), and Benthosaurus (Fig. 218), a small-eyed fish, with large ventral.

The pectoral rays of Bathypterois are strangely modified; the anterior ray is independent of the others, and so articulated that

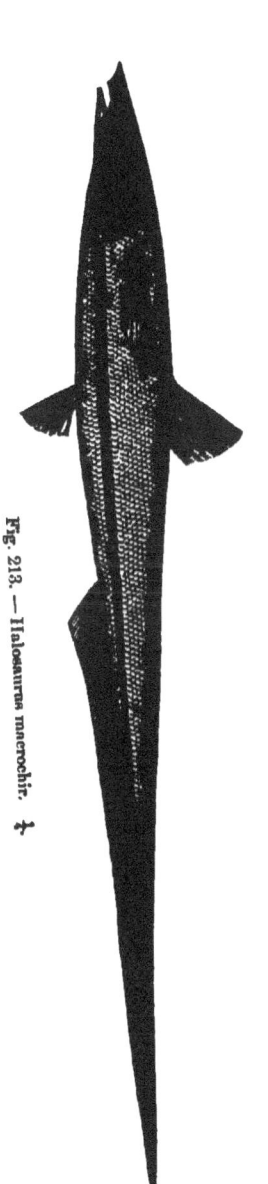

Fig. 213. — Halosaurus macrochir.

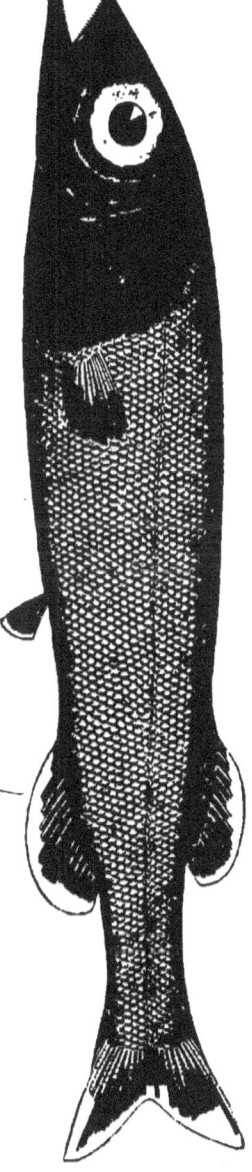

Fig. 212. — Alepocephalus Agassizii.

Fig. 214.— Chauliodus Sloani. ⅓. (U. S. F. C.)

Fig. 210.—Bathysaurus Agassizii. About ⅓.

Fig. 218. — Benthosaurus grallator. ⅓.

it may be extended in front of the head and used as an organ of exploration, so that we may imagine this fish feeling its way in the dark, and exploring the ooze to discover buried in it the animal upon which it feeds.

To the "pelagic Isospondyli" belong those groups which, like the Scopelidæ, are found from time to time at the surface, living or dead, and which, there is reason to believe, inhabit the intermediate depths of the ocean, having the power of ascending and descending developed to an extent which is not at present understood.

Among the deep-water groups named above occur the most abnormal specializations, such as powerful jaws, lancet-like teeth, prolonged tactile appendages, and enlargement of the tube-bearing scales. They have not the cavernous and feeble skeletons peculiar to the deep-sea gadoids, and many other families, which may have found their way gradually into deep water; they are, as a rule, compactly built, muscular, and are the most actively predaceous of the abyssal forms.

The pelagic groups do not, as a rule, exhibit special modifications of form, but they are, with few exceptions, provided with peculiar luminous appendages, which, like the cavernous skeletons and exaggerated mucous systems, have been by many writers attributed to deep-sea fishes in general.

In his "Challenger" letters, Willemoes-Suhm speaks of the luminosity of Scopelus. (Fig. 219.) It is well known to the fishermen of the Mediterranean that at the death of the fish the luminosity ceases. We frequently brought in scopelids in our tow-nets, and could observe the phosphorescence of the luminous spots, so arranged that it seems as if the anterior ones were intended to explore the regions in front of the fish, while those of the belly illuminated the water

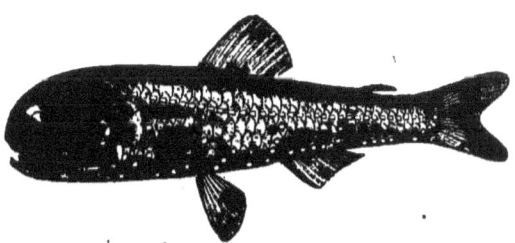

Fig. 219. — Scopelus Mülleri. ¼. (U. S. F. C.)

beneath it. The "Bombay duck," so common at certain periods in the Indian Ocean, belongs to this group of phosphorescent fishes. It is probably, with Scopelus, an inhabitant of deep water, coming to the surface only at certain times.

We may imagine some deep-sea types, when in search of their food, illuminating the water around them to a certain extent by their feeble phosphorescent light. Others carry beacons or specialized plates on certain parts of the head; others are resplendent with phosphorescent spots extending along the sides of the body, or the back, or ventral surface; while in others, again, long tactile appendages play the part of lights sent out to illuminate dark corners, or the fins themselves may be intensely luminous. Sometimes the whole body is phosphorescent, and diffuses a subdued light, as is the case with some of the deep-sea sharks. It is hoped that future investigations will solve for us the question whether all these phosphorescent fishes are not to a greater or less extent in the habit of swimming far from the bottom.

Ipnops is evidently a dweller on the bottom. The eyes of this fish have been carefully examined by Professor Moseley. They were at first considered phosphorescent organs, but they show a flattened cornea extending along the median line of the snout, with a large retina composed of peculiar rods, which form a complicated apparatus, destined undoubtedly to produce an image and to receive especial luminous rays.[1]

Malacosteus is the sole representative of a peculiar family, the affinities of which have never been defined. *Malacosteus niger*

---

[1] The existence of well-developed eyes among fishes destined to live in the dark abysses of the ocean seems at first contradictory; but we must remember that these denizens of the deep are immigrants from the shore and from the surface. In some cases the eyes have not been specially modified, but in others there have been modifications of a luminous mucous membrane, leading on the one hand to phosphorescent organs more or less specialized, or on the other to such remarkable structures as the eyes of Ipnops, intermediate between true eyes and specialized phosphorescent plates. In fishes that have been blinded and retain for their guidance only the general sensibility of the integuments and of the lateral line, these parts soon acquire a very great delicacy. The same is the case with tactile organs, and experiments show that barbels may become organs of touch adapted to aquatic life, sensitive to the faintest movements or the slightest displacement, with power to give the blinded fishes full cognizance of the state of the medium in which they live.

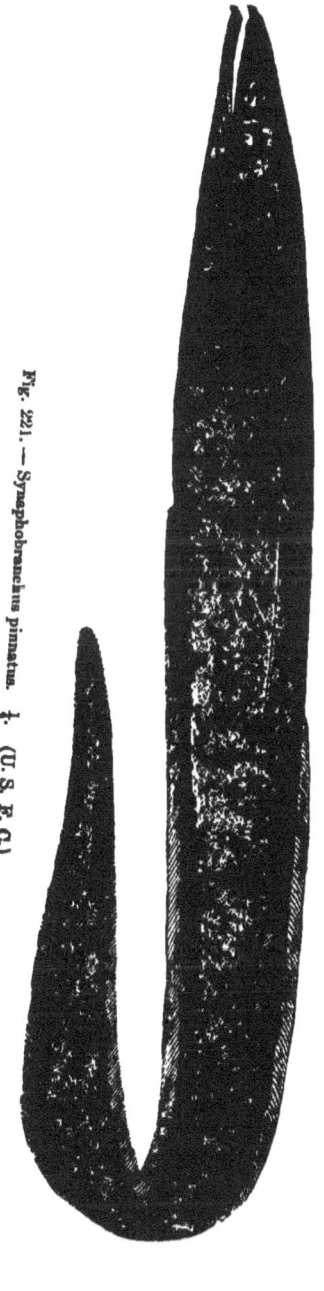

Fig. 221.—Synaphobranchus pinnatus. ⅓. (U. S. F. C.)

Fig. 222. — *Nemichthys scolopaceus*. ⅓. (U. S. F. C.)

Fig. 223. — Nettiatoma procerum. ⅟.

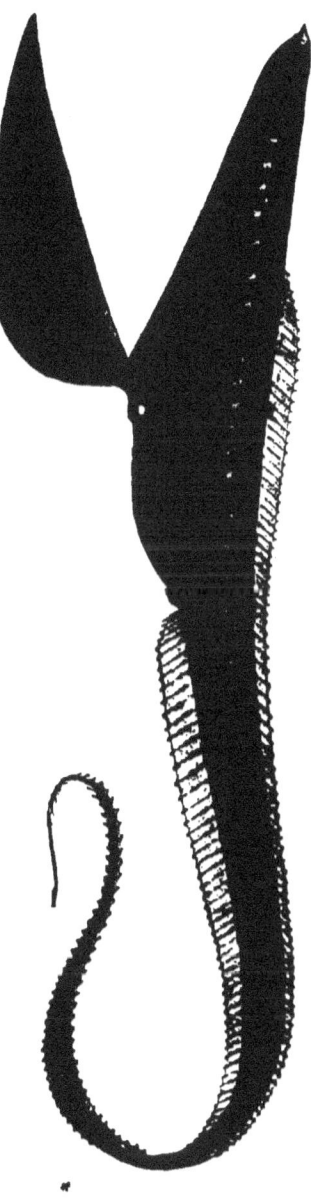

Fig. 224.—Gastrostomus Bairdii. ♀. (U. S. F. C.)

(Fig. 220) has been taken at the surface (dead), and also in the trawl at various depths from 335 to 1,000 fathoms, by the "Blake," "Albatross," and "Talisman." It has a luminous

Fig. 220. — Malacosteus niger. ⅓.

body under the eyes, and is possibly a form belonging to the intermediate depths of the ocean.

Characteristically abyssal is a familiar fish of our own coast, *Synaphobranchus pinnatus* (Fig. 221), ranging from 239 to 1,200 fathoms. Next come the Nemichthyidæ, popularly called the "snipe eels," exceedingly elongate, feebly finned forms, with the jaws prolonged and bill-like. *Nemichthys scolopaceus* (Fig. 222) occurs along our coast in 306 to 1,047 fathoms. Another typical genus living in considerable depths is Nettastoma, represented by *Nettastoma procerum* (Fig. 223), a new species taken by the "Blake" in 178 to 955 fathoms.

Some of the deep-sea fishes must find it most difficult to supply themselves with food. Such types as the astonishing Eurypharynx, discovered by the "Talisman," and its American ally, *Gastrostomus Bairdii* (Fig. 224), seem to meet the problem of foraging by a policy of masterly inactivity. Water and the food it contains pour into the mouth and the enormous cavity behind it, which is formed both above and below by the lateral folds of the head and of the anterior part of the body, constituting a huge pouch, capable of great expansion. The head thus becomes an immense funnel, the body of the fish being its shank. Perhaps the process of digestion is carried on in part in this pouch.

This fish undoubtedly lives in the soft ooze of the bottom, its head alone protruding, ready to ingulf any approaching prey. Its fins are atrophied, and the power of locomotion of this strange animal must be reduced to a minimum. The structure of the lateral line as described by Ryder is unique. There are groups of four and five stalked organs, more or less cup-shaped, the surrounding skin deeply pigmented. The function of these side organs is probably tactile, or they may serve some special purpose at the great depth at which these fish live. Analogous organs have been described in the head of the blind cave fish. It may be that the side organs are phosphorescent, like those of the scopelids. These side organs also recall the sense organs of embryo fish. The respiratory apparatus is unique among bony fishes. There are air-breathing slits, and the water which enters the buccal cavity escapes by a small opening in front of the rudimentary pectorals. The "Blake" took specimens of this fish in 898 fathoms. It also occurs between 389 and 1,467 fathoms.

Of the selachians, few representatives have as yet been brought to light by deep-sea explorers, nor is it to be expected that such large forms should be captured by the methods hitherto employed, although, as has been stated, a regular fishery for deep-sea sharks (Centrophorus) has existed from time immemorial off the coast of Portugal. A species of skate was taken by the "Blake" in 233–333 fathoms. Scyllium and Spinax also occur below 200 fathoms (*Centroscyllium Fabricii* down to 671). Only three species of selachians at all specialized for deep-sea life have as yet been found, unless perhaps we except Chlamydoselachus, the frilled shark, a representative of the devonian selachians, which is found off Japan, where it probably is an inhabitant of deep water. This is one of those interesting persistent types, like the Australian Ceratodus and the American ganoids: the gar-pike and mud-fish. The Japanese shark has the teeth of an ancient devonian type, and the embryonic characters of the lowest orders of recent sharks.

The lamper eel (*Petromyzon marinus*) and hag (*Myxine glutinosa*) have both been dredged below 500 fathoms.

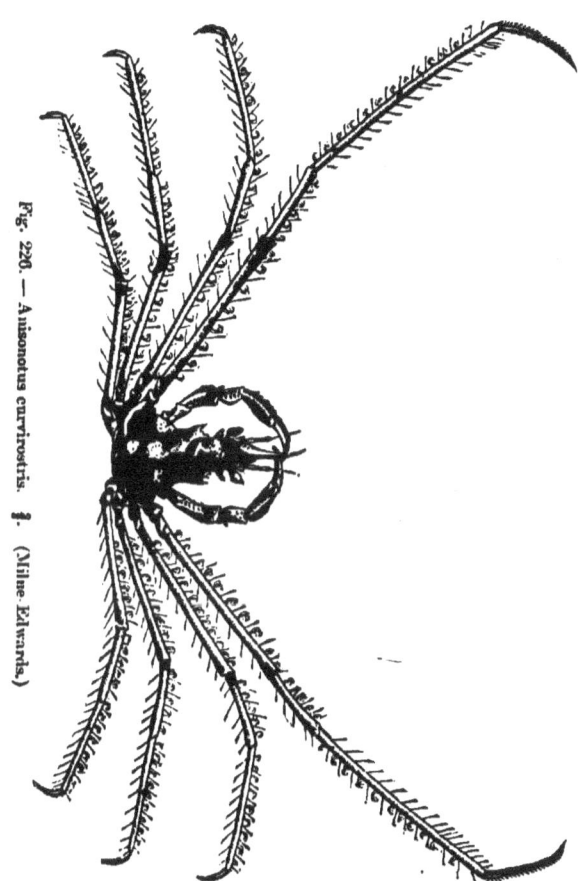

Fig. 220. — Anisonotus curvirostris. ♂. (Milne-Edwards.)

XVI.

CHARACTERISTIC DEEP-SEA TYPES. — CRUSTACEA.[1]

In a rapid survey of the "Blake" collections for the sake of noting some of the more interesting discoveries, the large number of very small and exceedingly long-legged spider-crabs (Maioidea) first attract attention. Species of this general character, such as *Anomalopus frontalis* (Fig. 225) and *Anisonotus*

Fig. 225. — Anomalopus frontalis. ²⁺⅕. (Alph. Milne-Edwards.)

curvirostris (Fig. 226), are found to be numerous, and many of them very abundant, at depths between 30 and 300 fathoms, in the West Indian region, and a few species extend northward to the south coast of New England. *Pisolambrus nitidus* (Fig. 227) represents another group of Maioidea inhabiting similar depths.

Among the Cancroidea (crabs and their allies), which are so

[1] Prof. Sidney I. Smith has kindly assisted me in preparing the account of the crustaceans.

characteristic of our littoral fauna, and are also found pelagic in the gulf-weed, there are comparatively few deep-water species and not so many novelties; but there are new species of a group of very small crabs, like Pilumnus, Neopanope, and Micropanope (Fig. 228), characteristic of the West Indian fauna at moderate depths. Off the Atlantic coast of the United States, however, *Geryon quinquedens*, previously known only from small specimens taken off the northern coast of New England, was found growing to enormous size at depths of from 200 to 800 fathoms, from the south coast of New England to points far south of Cape Hatteras. Specimens taken by the "Blake" show this species to be one of the very largest of the Brachyura, the carapace in some specimens being five inches long by six broad. Most interesting among the Leucosoidea is *Acanthocarpus bispinosus*. (Fig. 229.) Heretofore the only species of the genus known was *A. Alexandri*, which is armed with an enormous spine upon the outside edge of the claw, instead of on the side of the carapace. The claws are provided with a stridulating apparatus, which is rubbed against the edge of the carapace.

Fig. 227. — Pisolambrus nitidus. $\tfrac{2}{1}$. (Milne-Edwards.)

Fig. 228. — Micropanope pugilator. $\tfrac{1}{1}$$\tfrac{5}{}$. (Milne-Edwards.)

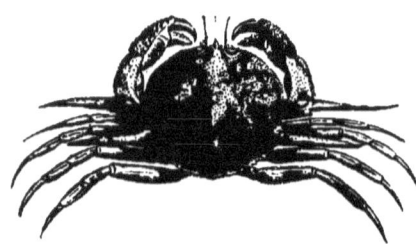

Fig. 229. — Acanthocarpus bispinosus. $\tfrac{1}{2}$. (Milne-Edwards.)

Quite striking is the large number of new forms of Dorippidoidea, a group previously unknown from the Western Atlantic and new to America. *Cyclodorippe nitida* (Fig. 230), a small species with smoothly rounded (Fig. 231) and highly polished carapace, will serve as an example. This and two other species of the same genus were taken in 90 to 300 fathoms. Belonging to the same group is the remarkable and

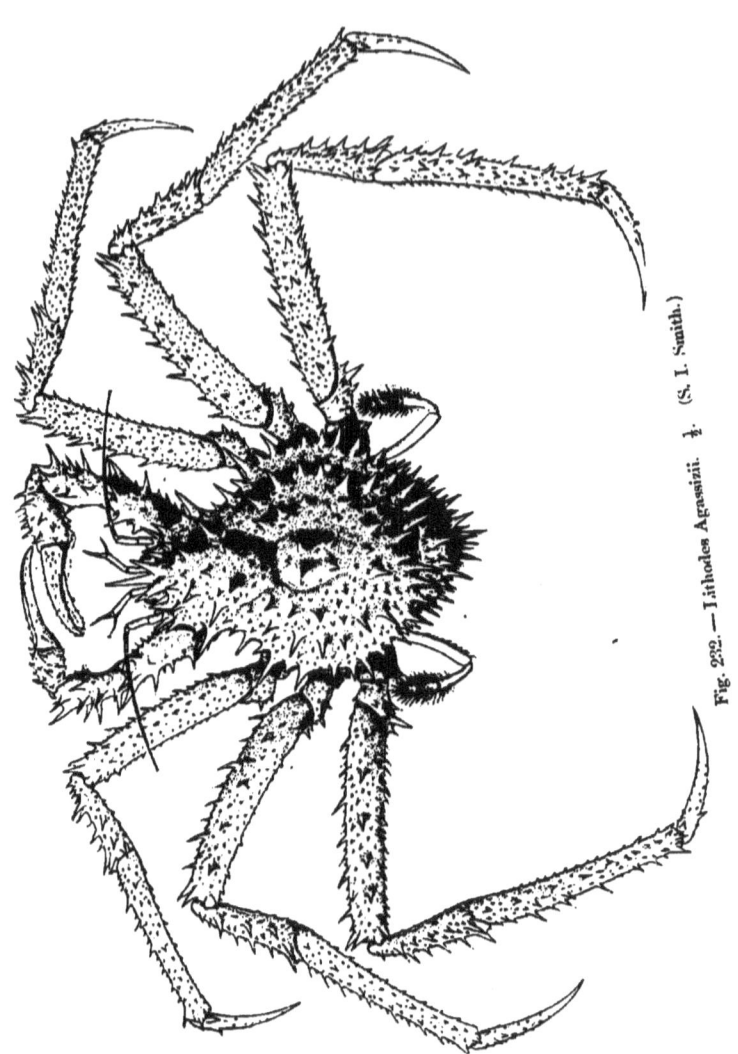

Fig. 232.—Lithodes Agassizii. ½. (S. I. Smith.)

little-known genus Cymopolia, of which no less than eight species are recorded from depths varying from 50 to 300 fathoms.

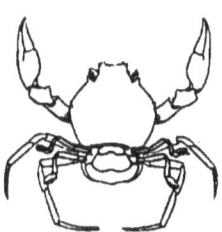

Fig. 230. — Cyclodorippe nitida. ⅘. (S. I. Smith.)

Cymonomus quadratus and *Cymopolus asper* represent two new species with the carapace projecting in a sharp rostrum in front. The latter species, taken in 75 to 150 fathoms, has normally developed eyes, while in the former,

Fig. 231. — Cyclodorippe nitida. ⅘. (Smith.)

taken in 200 to 500 fathoms, the eye-stalks are immobile spiny rods tapering to obtuse tips without visual elements; so that we may trace here, as it were, the mode of disappearance of the eyes in different groups of crustaceans.[1] The most remarkable species referred to this group is *Corycodus bullatus*, of which an imperfect specimen was taken between 175 and 250 fathoms. It has a somewhat pentagonal thick and very swollen carapace, covered with flattened tubercles resembling small rods.

Among the great number of new forms of Anomura (crustaceans intermediate between crabs and lobsters), none is more striking than the great spiny *Lithodes Agassizii*. (Fig. 232.) It is of a light pink color. Specimens have been taken with the carapace nearly seven inches long and more than six inches broad, and with the outstretched legs over three feet in extent. The whole integument of this magnificent species is very smooth, but the spines upon the carapace and legs are of needle-like sharpness, so that the greatest care is needful to handle even dead specimens without wounding the hands. Considering the pugnacious habits of crabs, it must be a formidable enemy among the members of its class. The spines are greatly elongated and very slender in young specimens, giving them an appearance very unlike that of the adult. This species was taken in 450 to 800 fathoms; it extends from the southern coast of New England to that of South Carolina.[2]

---

[1] In the Pycnogonidæ the shallow-water species have four eyes; the deep-water species none, or only rudimentary ones.

[2] Arctic species and genera were found by the "Blake" far south of their supposed range; the genus Lithodes was

Acanthodromia, which recalls from the shape of its carapace fossil crustacea characteristic of the secondary formation, and Dicranodromia, are peculiar new genera of Dromidæ inhabiting depths of 100 to 200 fathoms; while Homolopsis, with eyes nearly atrophied, is, like Cymonomus just mentioned, a Mediterranean genus which has been found by the "Blake" in the depths of the Caribbean. Homalodromia, a genus of the family of Homolidæ, is in some respects intermediate between it and the Dromidæ, two families thus far most distinct, and occurs in greater depths, from 300 to 600 fathoms.

Among the hermit-crabs (Paguroidea) the species thus far known were very similar, the head and claws alone being hard and calcareous, while the soft terminal parts of the abdomen are in the littoral species tucked away for protection into all

Fig. 233.   Fig. 234.
Xylopagurus rectus. ⅓.
(Milne-Edwards.)

sorts of bodies, such as shells and the like. It must be most difficult often for the deep-water species to find appropriate hiding-places, and it is not astonishing that the dredgings of the "Blake" have brought to light a number of remarkable new forms, whose characteristics unite them with the Macrura; as, for instance, *Pylocheles Agassizii*, which has a perfectly symmetrical tail. It lives in cavities excavated in fragments of stone formed of agglutinated sand. It entirely fills the cavity, closing the opening with the claws, which form a perfect operculum. *Xylopagurus rectus* (Fig. 233), a slender hermit-crab, inhabits tubes excavated in bits of wood (Fig. 234) or the hollow stems of plants open

previously known only from the Northern and Southern oceans. On the other hand, species previously known only from the West Indian region were discovered off the New England coast.

at both ends. To adapt it to its peculiar dwelling, the posterior rings of the tail are formed into a large and bilaterally symmetrical operculum of calcified plates, which closes the posterior opening as effectively as the stout claw does the anterior. The animal is straight, and has not the curved abdomen of the hermit-crabs; it enters its abode, not backwards, as do the hermit-crabs, but forwards, head first. *Mixtopagurus paradoxus* has a slightly asymmetrical tail, in which the rings are more or less distinct, but not completely calcified, so that it is intermediate in this respect between Pylocheles and the typical hermit-crabs. All three of these remarkable forms were taken in 100 to 200 fathoms in the West Indian region.

The species of Catapagurus inhabit depths of 50 to 300 fathoms from the southern coast of New England to the West Indies, and live in a great variety of houses which only imperfectly cover the animal, of which the front portion of the carapace is indurated. They are often associated with a colony of polyps, Epizoanthus (Fig. 235), or the house is built up by the base of a simple polyp, Adamsia, which has expanded laterally and united below so as to enclose the crab in a broad cavity. (Fig. 236.) The houses are generally built upon fragments of pteropod shells or worm-tubes as a nucleus. This is frequently resorbed.

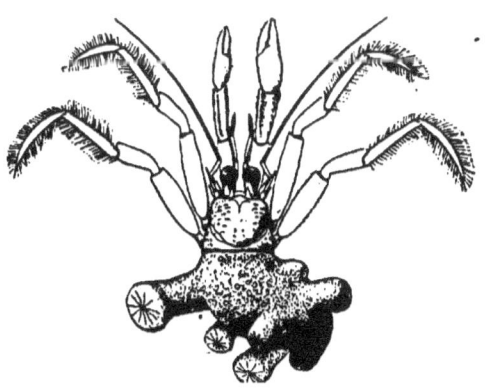

Fig. 235. — Catapagurus Sharreri. ⅔. (S. I. Smith.)

The Epizoanthus houses are very often disproportionately large for the crabs inhabiting them, having grown out on either side until they are several times broader than long. In spite of these enormous houses, both species of the genus probably swim about by means of the ciliated fringes of the ambulatory legs. A similar coöperative association between a sea-anemone and a crab from shallow water was already known, the polyp deriving most of its food from the remnants left by the crab, and the latter in its turn being hidden by the Actinia while creeping towards its prey.

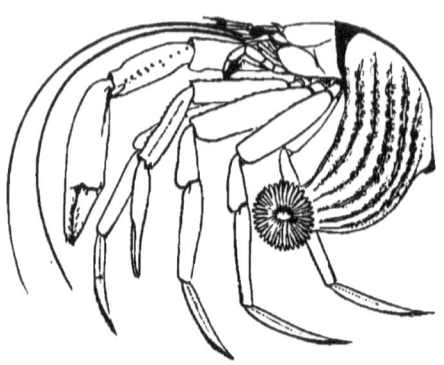

Fig. 236. — Catapagurus Sharreri. ♀. (S. I. Smith.)

Ostraconotus spatulipes, dredged from a little over 100 fathoms, is apparently the most aberrant of all the hermit-crabs. It appears to live without a house; the carapace is flexible, and resembles that of the Galatheoidea; the tail is so rudimentary that the bunches of eggs are supported by the feet.

The large number of Galatheoidea discovered is another prominent feature of the "Blake" collection. They were previously represented in our fauna by one imperfectly known species. They are very characteristic of deep water in depths of from 300 to more than 2,000 fathoms. This group of species is well illustrated by *Munidopsis rostrata*. (Fig. 237.) Some of the Galatheoidea have enormously long legs, with which to hunt for their prey in deep mud or in hidden corners, Munida. (Fig. 238.) Some of the small and weak forms of the group, Diptychus, are exceedingly abundant in 100 to 700 fathoms among the branches of gorgonians, and others in the interior of some of the delicate siliceous sponges; they appear greatly disturbed, running in all directions, when brought to the surface.

None of the deep-water Macrura have attracted more notice than the Eryonidæ, or "Willemœsia group of crustacea," first brought into prominent notice by the "Challenger" expedition. No less than five new species of this group were discovered at depths ranging from 100 to 1,900 fathoms; they are admirably illustrated by *Pentacheles sculptus*. (Fig. 239.) The eyes are sessile and peculiarly modified in all the species. In *Pentacheles sculptus* the eyes, or ophthalmic lobes rather, com-

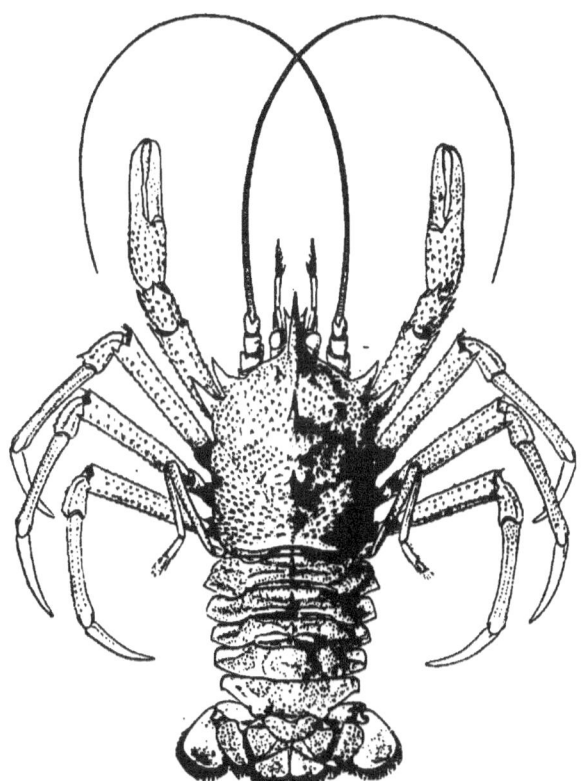

Fig. 237. — Munidopsis rostrata. ¼. (S. I. Smith.)

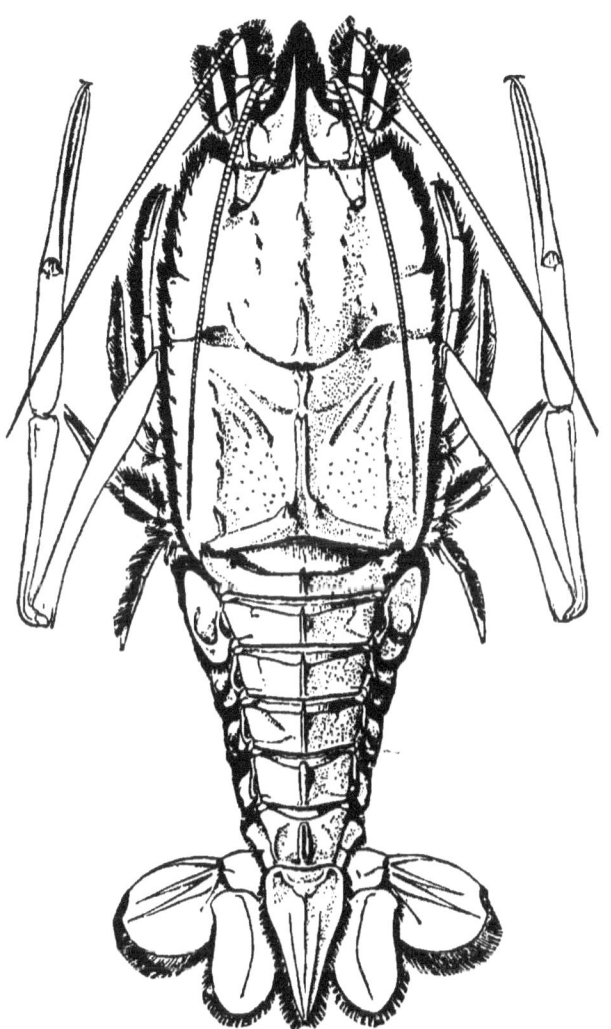

Fig. 239. — Pentacheles sculptus. ¼. (S. I. Smith.)

pletely fill deep orbital sinuses in the front of the carapace in which they are imbedded. The Willemœsiæ have a very wide geographical distribution, and they are peculiarly adapted for burrowing in soft ooze, in which they seem to live. Some of the species are wonderfully transparent. They are the repre-

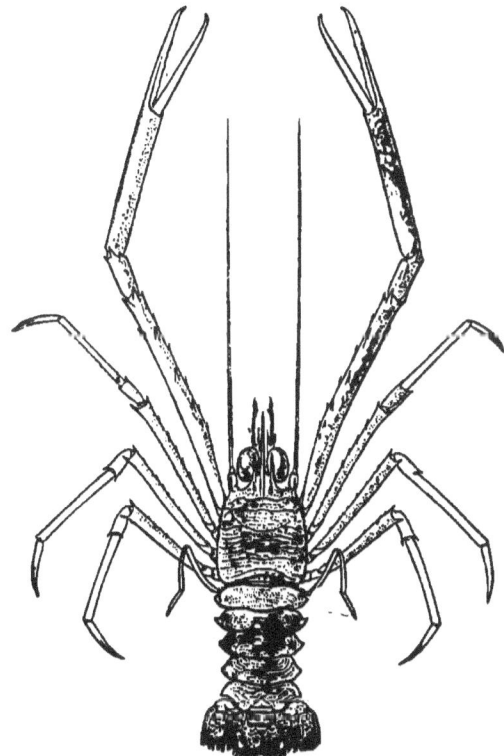

Fig. 238. — Munida. ¼. (S. I. Smith.)

sentatives in our seas of the fossil Eryonidæ, which flourished in the jurassic lithographic beds of Solenhofen, in Bavaria. It is interesting to note that the eyes of the fossil species were extraordinarily developed.

Nephropsis Agassizii (Fig. 240), the only species of Astacidea discovered, belongs to a genus previously known only from a single imperfect specimen dredged in the Bay of Bengal. The genus is closely allied to our lobster: its species have very small and colorless eyes.

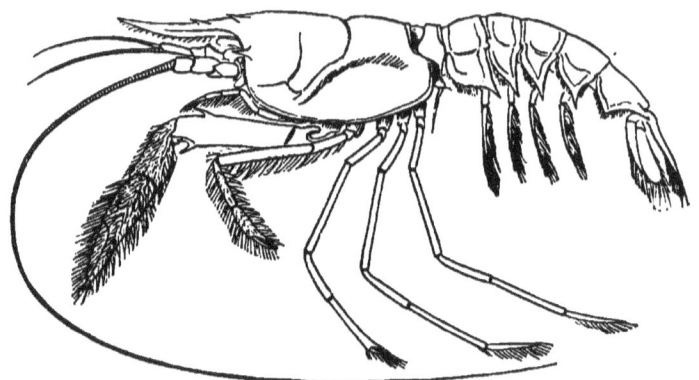

Fig. 240. — Nephropsis Agassizii. ♂. (S. I. Smith.)

Phoberus cœcus (Fig. 241), taken in 416 fathoms off Grenada, is a gigantic crustacean, combining, according to Milne-Edwards, characters of several. families of macrurans. It is as large as a lobster, the carapace in one specimen being seven inches in length; and the whole animal, from the end of the tail to the tip of the outstretched claws, is twenty-eight inches, while the claw alone is eight inches. The eyes are rudimentary, and do not project beyond the carapace.

It is difficult to draw any conclusions from the great diversity presented by the conditions of the organs of sight in the crustaceans. Even among allied species we find that some are blind, while others have well-developed organs of vision; in one group the eyestalks are flexible, while they are rigid in the next. One cannot help being struck with the fact that a comparatively small number of deep-sea crustaceans have lost their eyes.

Glyphocrangon (Fig. 242) represents a new family, of which several species were taken both in the West Indian region and off the Atlantic coast of the United States in 250 to 1,200 fathoms; these very characteristic deep-water forms are all large and shrimp-like, with massive, highly sculptured, spiny, and tuberculose integument. The carapace, owing to a peculiar articula-

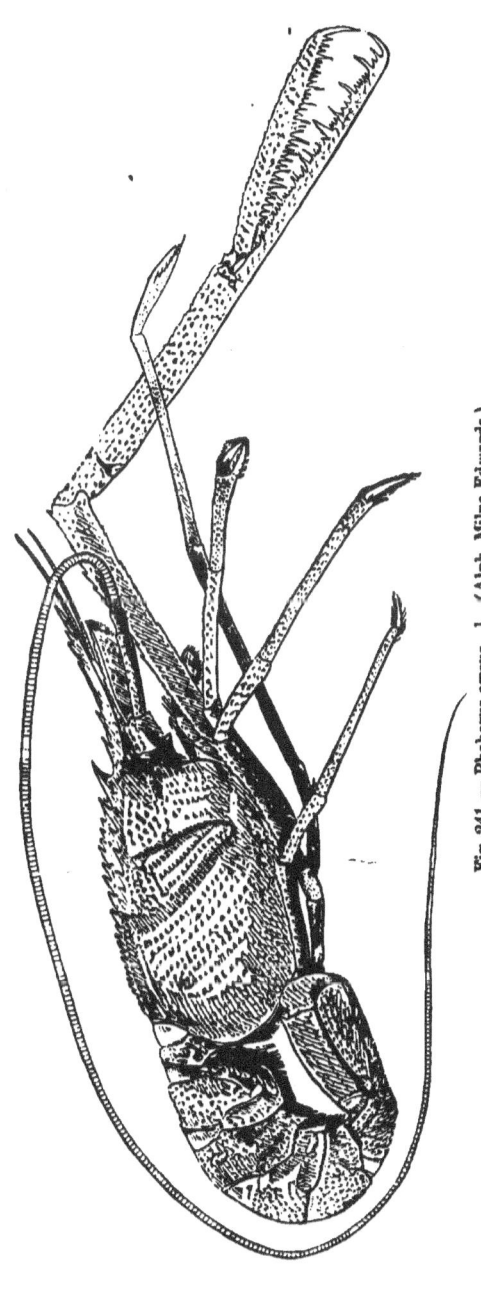

Fig. 241. — Phoberus cæcus. ½. (Alph. Milne-Edwards.)

tion formed by a projection of its margin and by processes of the external feet-jaws, is capable of a slight motion, a character unknown among decapods. The hinges of the last three articulations of the rings of the tail are modified, so that they can be

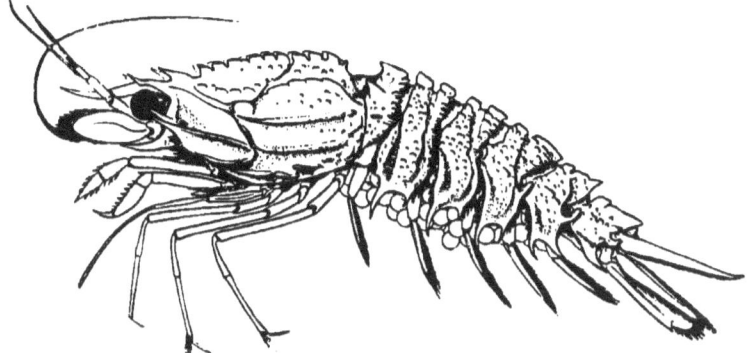

Fig. 242. — Glyphocrangon aculeatus. ½. (S. I. Smith.)

clamped, and the animal can hold the terminal rings firmly extended as a means of self-defence.

Sabinea princeps (Fig. 243), taken in 400 to 700 fathoms off the Atlantic coast of the United States, and a closely allied spe-

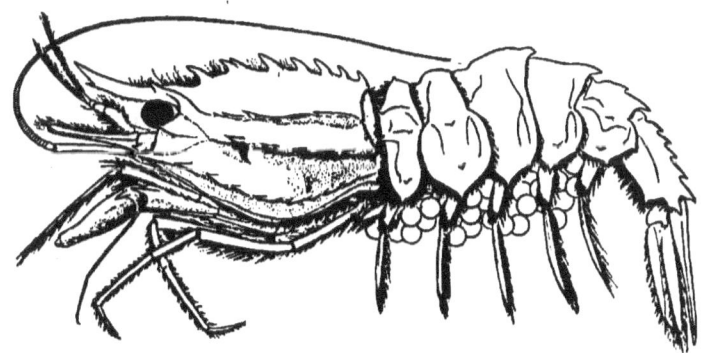

Fig. 243. — Sabinea princeps. ½. (S. I. Smith.)

cies from off Guadeloupe, are the largest known species of the family of Crangonidæ, and many times larger than the two northern species of the genus. *S. princeps* reaches a length of five inches or more.

Numerous new species of Pandalus, some of them very large and with greatly elongated legs, and of the allied genus Heterocarpus (Fig. 244), in which the carapace is beautifully carinated,

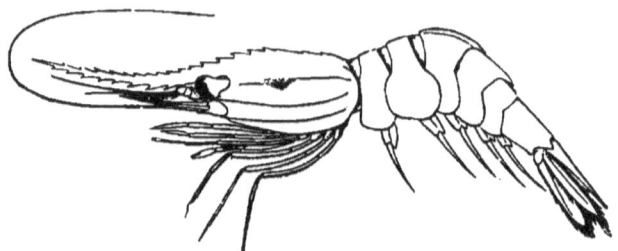

Fig. 244. — Heterocarpus carinatus. ⚥. (S. I. Smith.)

were taken in 200 to 1,000 fathoms; they are apparently characteristic of the fauna at that depth in the West Indian region. The species of the new genus Stylodactylus, dredged from 400 to 500 fathoms, probably represents a new family of Caridea.

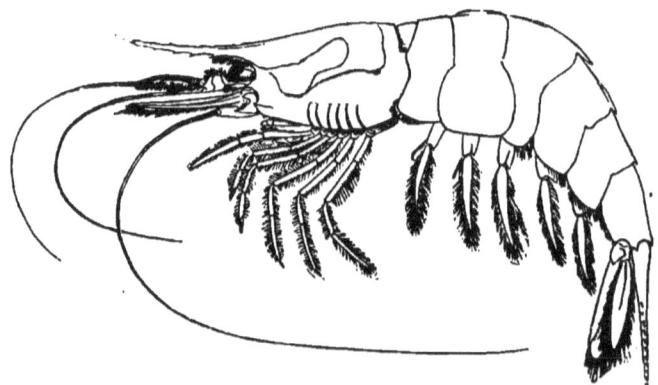

Fig. 246. — Acanthephyra Agassizii. ½. (S. I. Smith.)

The oral appendages and branchiæ belong to a peculiar type of structure, and the claws of the first and second pairs of legs are very long and slender, with slender multiarticulate and hairy digits. *Nematocarcinus ensiferus* (Fig. 245), of a bright rose-color, from 800 to 1,400 fathoms, and *N. cursor*, from 500

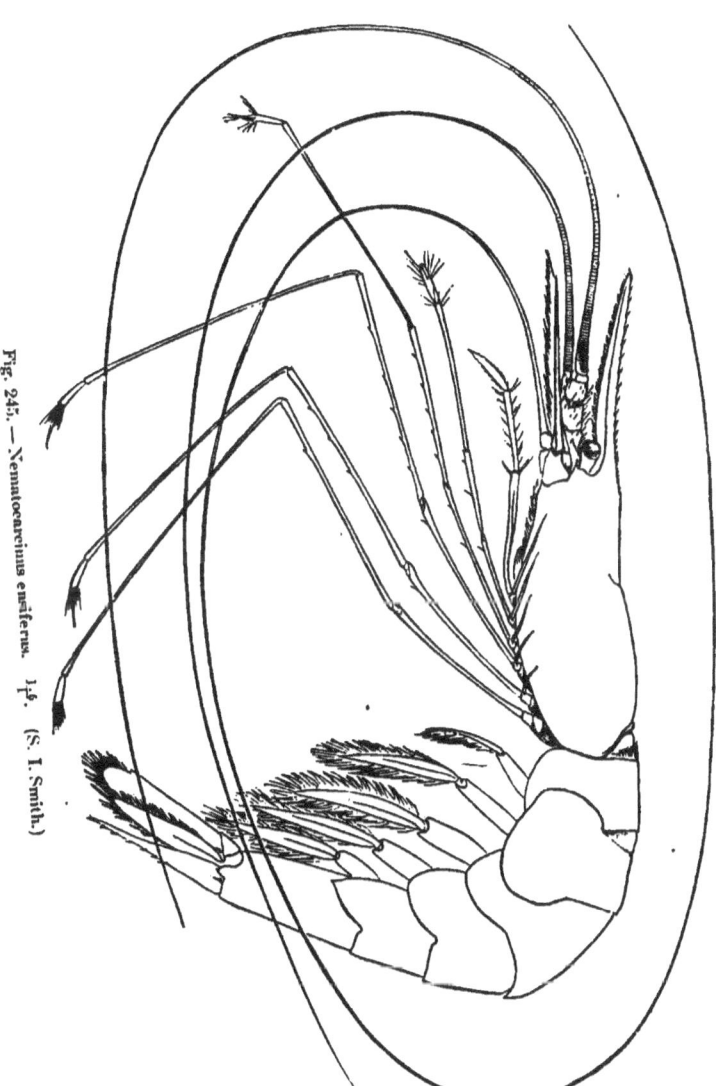

Fig. 245. — Nematocarcinus ensiferus. 1⅓. (S. I. Smith.)

CHARACTERISTIC DEEP-SEA TYPES. — CRUSTACEA. 47

fathoms, represent a new and very peculiar family, of which the species are often abundant in deep water. Their exceedingly long and very delicate legs, three to four times the length of the body, tipped with fascicles of long setæ, are apparently intended as an adaptation for resting on very soft oozy bottoms.

New species of the little known genus Oplophorus, and the new genera Acanthephyra (Fig. 246), Notostomus, and Meningodora (Fig. 247), make up a group of species of which almost nothing was known before the explorations of the "Blake," although they are very frequently taken in the trawl at great depths. The structure of the articular appendages of these species is very much like that of the schizopods and the majority of larval macrurans.

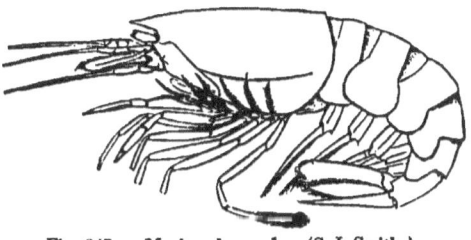

Fig. 247. — Meningodora. ¼. (S. I. Smith.)

Some of the species of Notostomus grow to a large size, are very deep crimson when first taken from the water, and are among the most striking of all the abyssal Caridea.

The only Penæidæ which have been as yet described are from

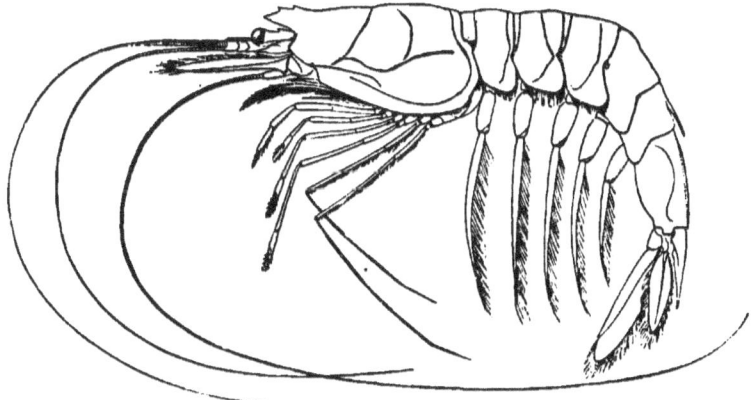

Fig. 248. — Benthœcetes Bartletti. ¼. (S. I. Smith.)

off the Atlantic coast of the United States. These, though few in number, are very interesting. *Benthœcetes Bartletti* (Fig. 248) will serve as an example. In this species the filaments of each antenna are greatly elongated, — fully once and a half the length of the body; the legs increase in length towards the posterior extremity, and the three anterior pairs have minute claws; the dactyli of the two posterior pairs, nearly twice as long as the preceding pair, are exceedingly weak and slender, and are evidently tactile rather than ambulatory organs, — modifications which seem adapted to the deep-sea life of these animals. We are constantly struck with the exquisite delicacy and great diversity of the organs of vision, of hearing, of touch, and even of smell, in the deep-water crustaceans. The antennæ and claws are frequently of excessive length, as if to facilitate exploration of the ooze and the sounding of objects.

We find in deep water huge schizopods, Gnathophausia (Fig. 249), of a beautiful red color. The majority of schizopods previously known were mainly pelagic, and belong to a group of small crustaceans which have the thoracic feet all alike, divided into two branches and sometimes carrying free gills. Some of these deep-water schizopods are provided with special organs of phosphorescence, such as luminous plates behind the eyes or over the legs. Among the various groups of crustaceans some have phosphorescent eyes, while in others the phosphorescence is diffused, or limited to special parts of the body at the time of breeding, or when irritated.

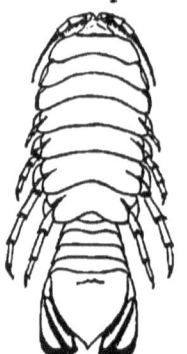

Fig. 250. — Syscenus infelix. About 1/5. (Harger.)

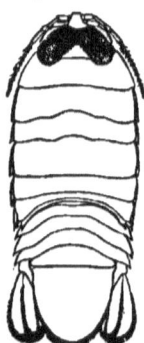

Fig. 251. — Rocinela oculata. ½. (Harger.)

Among the Atlantic species of isopods, we may figure the bright orange Syscenus (*S. infelix*, Fig. 250), which is found at a depth of nearly 400 fathoms, and Rocinela (*R. oculata*, Fig. 251), the upper surface of the head of which is nearly covered with large ocelli arranged in rows. From the collection made in the West Indian

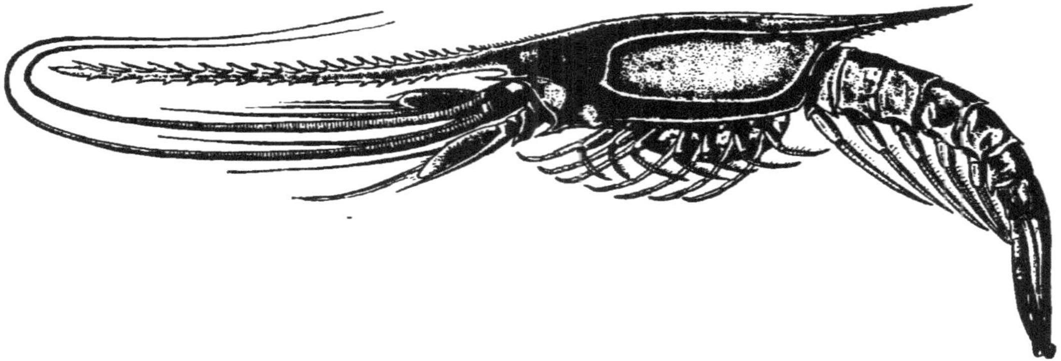

Fig. 240. — Gnathophausia Zoea. ♀. ⁷⁄₁. (A. Milne-Edwards.)

Fig. 252. — Bathynomus giganteus. ⅓. (A. Milne-Edwards.)

region only a single species, *Bathynomus giganteus* (Fig. 252), has been described, but this is by far the largest isopod known, and is more than eleven inches long! The eyes of this giant are placed on the lower side of the head, and consist, according to Milne-Edwards, of no less than four thousand facets.

The amphipods have not been studied, but the collection from the Atlantic coast of the United States contains several interesting species; among them the great angular and spiny *Epimeria loricata* (Fig. 253), first described from specimens taken by the Norwegian expedition in the North Atlantic, and a single specimen of the very peculiar *Neohela pasma*.

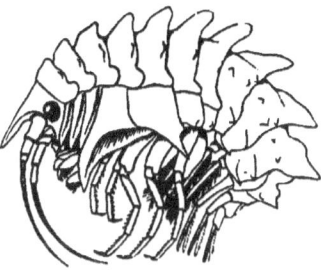

Fig. 253. — *Epimeria loricata.* $\frac{1}{1}$. (S. I. Smith.)

The pycnogonids from the West Indian region have not yet been described, but those from the Atlantic coast of the United States, which have been studied by Prof. E. B. Wilson, are especially interesting. The most striking

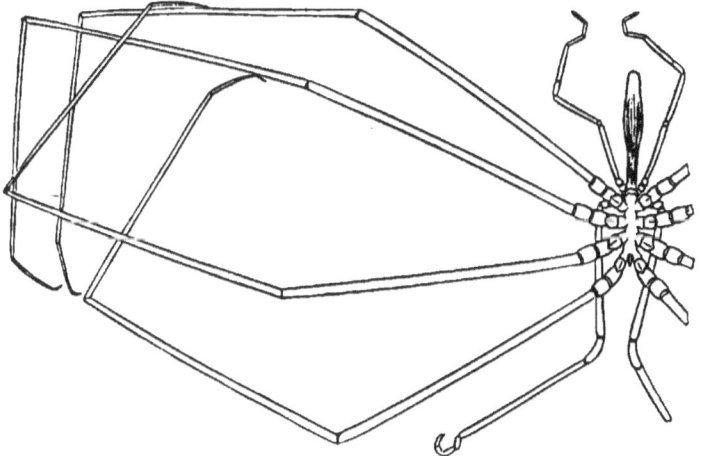

Fig. 254. — *Colossendeis colossea.* $\frac{2}{3}$. (E. B. Wilson.)

feature of the species is their great size, most of them being gigantic as compared with shallow-water species. There were ten species in the "Blake" collection, and half of them were new. The largest species is *Colossendeis colossea* (Fig. 254), in which the slender legs are nearly two feet in extent, and the rostrum more than an inch long, while the more slender *Colossendeis macerrima* spreads to fourteen inches, and has a rostrum fully as long as in the larger species. These species were taken in 500 to 1,200 fathoms. The new genus Scæorhynchus (Fig. 255) is remarkable for its spiny body and swollen and reflexed rostrum; the legs of *S. armatus* (Fig. 256), the single species taken below 1,200 fathoms, are nearly five inches in length. The most abundant species of Nymphon is also the largest known species of the genus. One of the species of the new genus Pallenopsis, dredged from 260 to 330 fathoms, is more than twice as large as any of the species from allied genera belonging near the shore or in comparatively shallow water.

There is a great contrast between the life of the communities of barnacles, such as we find living crowded on our rocks and floating on the surface, and that of the comparatively solitary deep-sea cirripeds Scalpellum, Verruca, and the like. This is readily understood when we remember that the living or dead organic matter floating on the surface in the wake of currents, and along the shores, supplies the former with a large amount of food, while the conditions of life at the bottom are far from favorable for the species living in deep water.

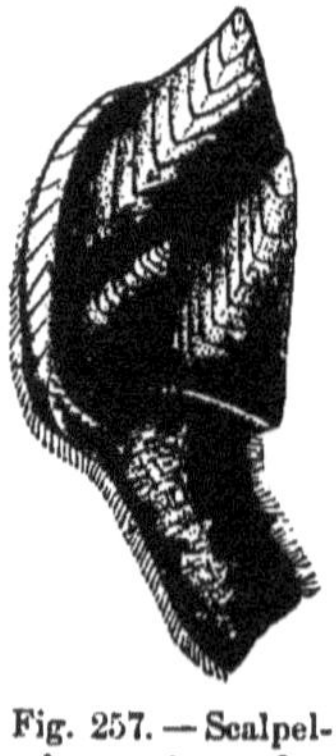

The abyssal cirripeds are usually attached to nodules, to dead or living shells, to corals, large crustaceans, spines of sea-urchins, and the like. *Scalpellum regium* (Fig. 257), a pedunculated form, first named by Wyville Thomson, is one of the largest species of the genus; it has been dredged by the "Challenger" from nearly 3,000 fathoms, and is quite common in the West Indies. *Verruca incerta* (Fig. 258) also is not an

Fig. 257. — Scalpellum regium. ¼. (Hoek.)

Fig. 258. — Verruca incerta. ⅝. (Hoek.)

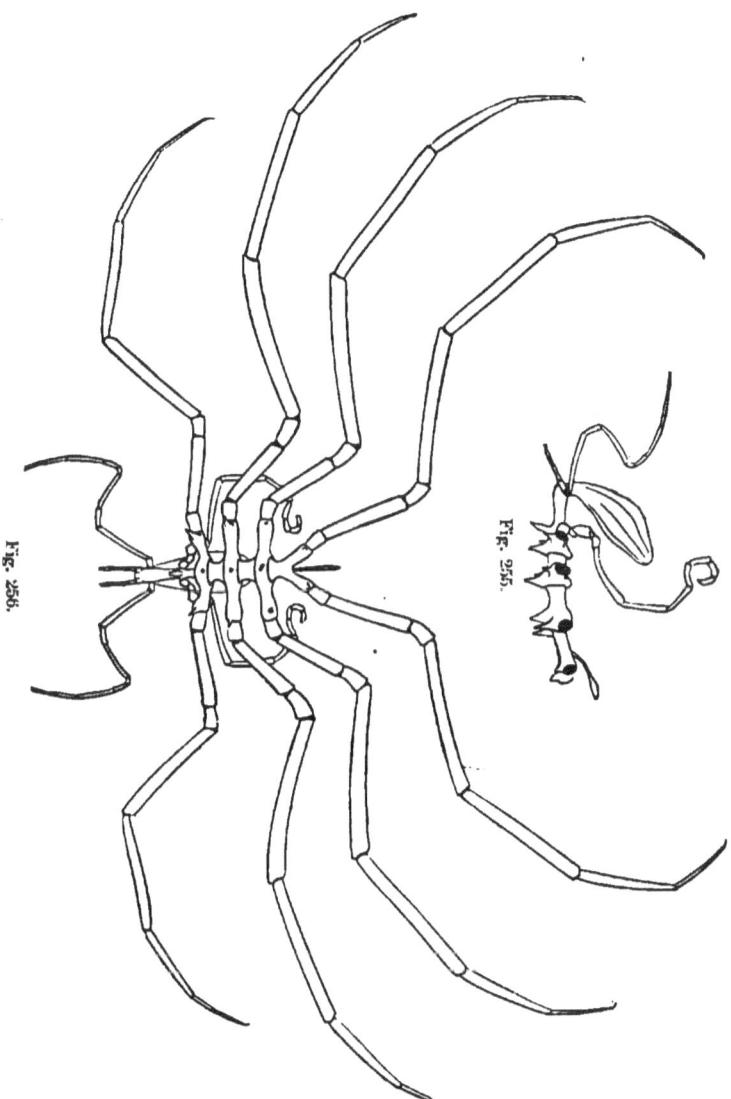

Figs. 255, 256. — Scæorhynchus armatus. $\frac{1}{1}$. (E. B. Wilson.)

uncommon West Indian type from the globigerina ooze: it belongs to the group having no peduncle.

As has been noticed by Hoek, the presence of Scalpellum and Verruca in the great depths of the ocean coincides in a striking manner with the palæontological history of these genera. They are found in the secondary deposits, yet the genus Pollicipes, another of the pedunculated cirripeds, dating back to the oölite, is only a littoral genus in our seas.

The ostracods are minute crustaceans, the dead tests of which occur in nearly all the bottom deposits. They are very abundant fossils, but the deep-sea dredgings have not as yet revealed any type of importance. Many of the ostracods (Fig. 259) are pelagic; only a comparatively small number live at any considerable depth; they are denizens of shallow water or of moderate depths.

Fig. 259.—Cypris. Greatly magnified.

XVII.

CHARACTERISTIC DEEP-SEA TYPES. — WORMS.[1]

THE collection of worms made by the "Blake" expeditions is remarkably rich, and not merely confirms in general the relations which similar materials from other deep-sea expeditions had already shown, but in a number of instances furnishes a most desirable supplement to the results of the earlier expeditions. Unfamiliar worms are here found in well-preserved specimens, while worm-cases which had before only been seen empty have been dredged occupied by their builders. Annelids make up the larger part of this collection, and among them the tubicolous annelids are by far the most numerous. One of the large Eunicidæ, *Hyalinœcia tubicola* (Fig. 260), was specially numerous; its tubes, sometimes fully fifteen inches in length, often filled the bottom of the trawl when it was dragging on muddy bottoms. Some of these genera are most striking from the exquisite beauty of their tubes, which are composed of siliceous spicules, and dead pteropod shells, and also from their strange association with corals, gorgonians, sponges, starfishes, mollusks, and ascidians. A species of Phorus was frequently accompanied by a large annelid, comfortably established in the axis of the shell, with its head close to the aperture. Of other worms the Nemertinæ are represented by isolated fragments; the gephyreans by Sternaspis, from a depth of 158 fathoms, and Aspidosiphon, from 190 fathoms; while many still undetermined species of Phascolosoma

Fig. 260. Hyalinœcia.

[1] The following account of the worms is taken from the Preliminary Report of Prof. Ernst Ehlers, of Göttingen, who has supervised the drawing of the figures.

extend from the littoral region as far as the greatest depth here recorded, one species having indeed been brought up in a Dentalium shell from a depth of 1,568 fathoms. Although so numerous, no new forms of these groups were collected either by the "Challenger" or "Blake," with the exception, perhaps, of some of the tubicolous types in deep water. Furthermore, these groups have but a slight significance as compared with the chætopods of the collection. The existence of chætopods in certain localities where the animals themselves are not found may be inferred by the presence of their tubes. Like the littoral species of Maldanidæ, Clymenæ, Serpulæ, and their allies, they must cover extensive tracts of ground with their tubes. Yet such a conclusion is not always admissible without further evidence; it can be accepted only when the individual worm builds his tube in so characteristic a way that there is no possibility of mistaking it for that of other annelids. Several times tubes which from their whole appearance have been taken for worm-cases were discovered to be inhabited by crustaceans (Amphipoda). We cannot always decide if the occupant of the tube was also its builder.[1] When no foreign material is used in the construction of the tube except mud consolidated by the secretions of the worm, the tubes of very different species of worms may have a great similarity among themselves; when, on the contrary, various foreign materials are cemented

Fig. 261.—Diopatra Eschrichtii. ⅓.   Fig. 262.—Diopatra glutinatrix.   Fig. 263.—Hyalopomatus Langerhansi. ⅓.

[1] Prof. S. I. Smith has observed the peculiar tubes in which some amphipods live; they are mainly built up of pellets of their excreta, cemented together by threads spun by the little crustacean.

in the tubes, such marked peculiarities may occur in their choice and application that from a fragment of the tube the builder can be inferred with certainty, and the form of
the tubes (Figs. 261, 262, 263) may even be so characteristic that there is no danger of mistaking them for other tubes. We have examples of this kind especially in the Eunicidæ, and also in the Maldanidæ (Fig. 264), Terebellidæ, Sabellidæ, and Serpulidæ. In determining the distribution of the worms, it must be remembered that uninhabited tubes, usually filled by mud or other material from the bottom, may be transported by currents.

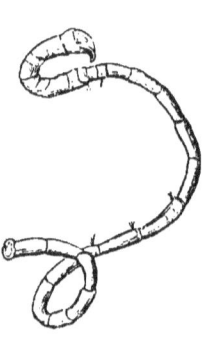

Fig. 264. — Maldane cuculligera. ⅔.

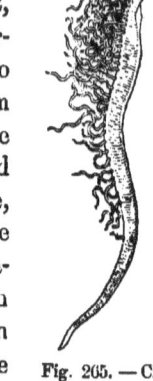

Fig. 265. — Cirratulus melanacanthus. ⅔.

Many of the principal types of the littoral annelids have not

Fig. 266. — Amphinome Pallasii. ⅔.

been dredged beyond the hundred-fathom line; such familiar groups as the Syllidæ, Nereidæ, Cirratulidæ (Fig. 265), and Amphinomidæ (Fig. 266), have no representatives at that depth, while the Phyllodocidæ, Ariciidæ, Terebellidæ, and Sabellidæ extend to 300 fathoms, and such families as the Polynoidæ (Fig. 267), Eunicidæ, Opheliidæ, Aphroditidæ, and Serpulidæ live beyond the five-hundred-fathom line, where occur also the Ampharetidæ, many of which live in tubes lined with a chitinous layer.

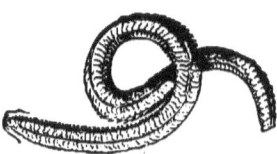

Fig. 267. — Sthenelais simplex. 1¼.

Of the families here enumerated, none has so important a bearing on the character of the faunal region as that of the Eunicidæ. Their representatives are found in far the greatest number of localities; they range from the littoral district to the lowest depths at which chætopods have been dredged by the "Blake." They are represented by the largest number of genera (Diopatra, Onuphis, Eunice, Rhamphobrachium (Fig. 268), Marphysa, Lisidice, Lumbriconereis, Arabella), and, judging from the large number of their tubes met with in many localities, they must form an essential part of the fauna. It is easily seen, however, that the various genera of this family show differences in their vertical range, the bearing of which will perhaps be more clearly understood when the conditions of temperature of their habitat are taken into account in connection with it. Thus the *Eunice conglomerans*, judging from the abundance of its paper-like irregular tubes (Fig. 269), is a characteristic inhabitant of the littoral belt, as far as 100 fathoms. From deeper waters come the tubes of the *Eunice tibiana* Pourt.; they descend to 243 fathoms, about to the region where the Eunicidea of the species Diopatra and Onuphis appear, some of which frequently build very peculiar tubes; such as the flat, parchment-like tubes with cemented sponge spicules of *Diopatra Pourtalesii*, and others mentioned by Pourtalès in his preliminary account of the results of his first expedition.

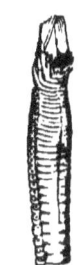

Fig. 268. Rhamphobrachium Agassizii. ⅔.

Fig. 269. — Eunice conglomerans. ⅔.

Among these chætopods species now appear which perhaps belong exclusively to the deep sea; they are separated from Diopatra-like forms, with large leaf-like expansions of the anterior appendages, and with long hook-like curved bristles at the point. The Diopatra (Fig. 270) group begins near the hundred-fathom line; it becomes particularly numerous at about

Fig. 270. — Diopatra glutinatrix. ⅖.

500 fathoms, and still has one representative at a depth of nearly 1,000 fathoms.

In connection with the important part here taken by the Eunicidæ in the faunal combination of a marine area, it is interesting to remark that among the annelids of the lithographic shales of Bavaria the Eunicidæ are those which, in various forms, are most richly represented.

One of the most interesting of the deep-water types collected

Fig. 271. — Buskiella abyssorum. ½. (McIntosh.)

by the "Challenger" is the eminently embryonic Buskiella (Fig. 271), which bears the closest resemblance to a chætopod larva.

Of other families found in deep water, the Polynoidæ and the Aphroditidæ may be especially mentioned. But as they never live in communities, and do not, as a rule, build large tubes, they are, like the Opheliidæ, less characteristic of the localities to which they belong than the Maldanidæ, or the Ampharetidæ; their large tubes, built of mud, and sometimes associated with those of the Eunicidæ, must, judging from the masses in which they are found, be a marked feature of certain localities.

CHARACTERISTIC DEEP-SEA TYPES. — WORMS. 57

It is interesting to find that the Serpulidæ (Fig. 272) also occur at great depths, because Ehlers, in working up the annelids of the "Porcupine" expedition, had noticed their absence in deep water, and left it undecided whether they were excluded by the peculiar nature of the bottom or by the low temperature of the deep sea. But it is not uncommon in the deep water of the Gulf of Mexico to bring up rocky fragments which, judging from the amount of mud brought up by the trawl at the same time, must form isolated patches, and in these undoubtedly the Serpulæ thrive. (Fig. 273.) Terebellidæ and Serpulidæ have been obtained by the "Challenger" at depths of nearly 3,000 fathoms. Of course, where the tubes are composed of secretions, as in Hyalinœcia, they are independent of their surroundings and of the character of the bottom. But the majority of the tube builders depend upon the material at their disposal, using, to strengthen their tubes, either sand, or mud, or larger solid particles, such as foraminifers, bivalves, sponge spicules, and the like.

Fig. 272. — Pomalostegus stellatus. ?/1.

Fig. 273. Hyalopomatus Langerhansi. 4/1.

XVIII.

CHARACTERISTIC DEEP-SEA TYPES. — MOLLUSKS.

CEPHALOPODS.

THE shoal-water species of cephalopods, the squids and cuttle-fishes, live upon the bottom; but, being powerful swimmers, they are capable of extensive migration, so that with them as with fishes it will always be difficult to ascertain the depth from

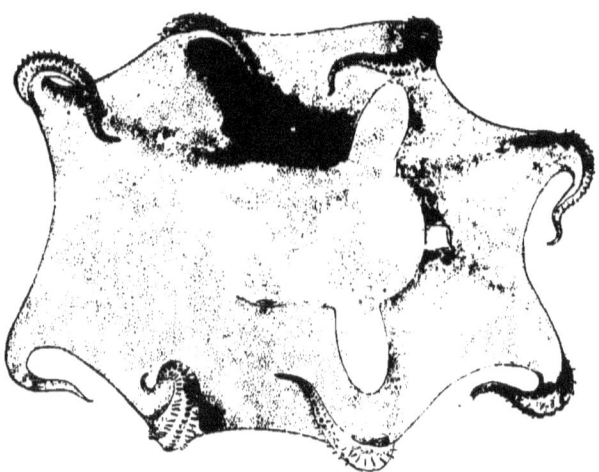

Fig. 274.—Opisthoteuthis Agassizii. Abt. ½. (Verrill.)

which they have been obtained. Many of them are pelagic, and serve as food for a large number of marine animals.[1]

Professor Verrill, who has examined the cephalopods collected by the "Blake," mentions as specially noteworthy the following: *Opisthoteuthis Agassizii* (Fig. 274), a species with a broad body of a dark chocolate color, long fins, and arms united

[1] Very common in the Gulf Stream is the *Sthenoteuthis Bartrami*, large specimens of which are often caught on the surface. It is known as the "flying squid," often darting out of the water in the velocity of its movements.

CHARACTERISTIC DEEP-SEA TYPES. — CEPHALOPODS.   59

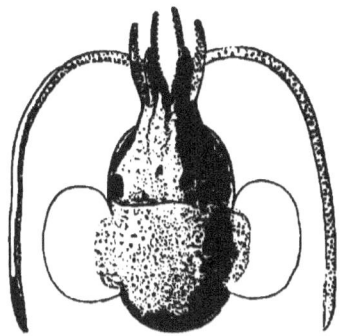

Fig. 275. — Nectoteuthis Pourtalesii. ⅓. (Verrill.)

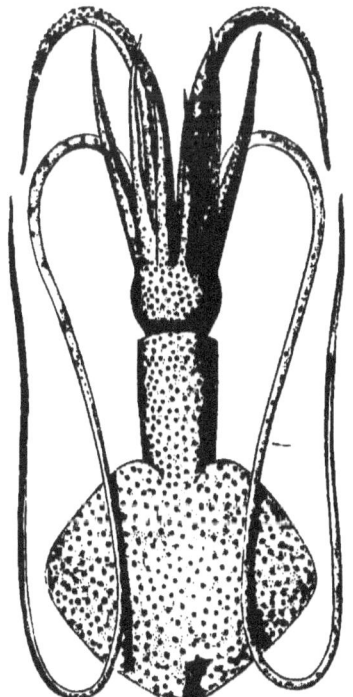

Fig. 276. — Mastigoteuthis Agassizii. ½. (Verrill.)

nearly to their tips by a thick soft web; among the cuttle-fishes, a small reddish-brown species, *Nectoteuthis Pourtalesii* (Fig. 275), characterized by its short thick body and the great size of its ventral shield; and the remarkable genus Mastigoteuthis (Fig. 276), the type of a new family, with very unequal arms, and a large caudal fin, of an orange-brown color, occupying about half the length of the body.

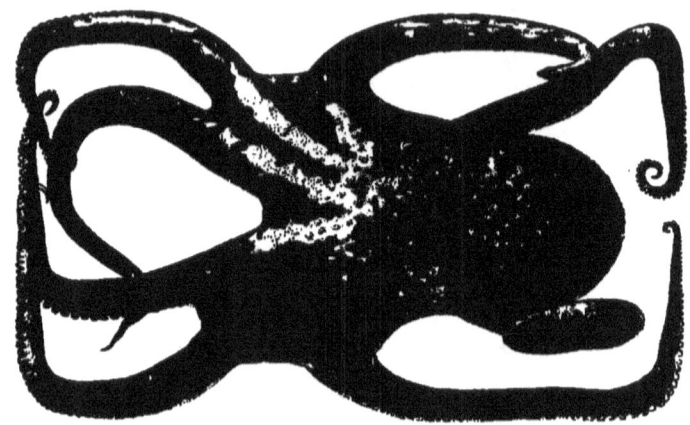

Fig. 277. — Eledone verrucosa. ½. (Verrill.)

A stout species of octopoid, *Eledone verrucosa* (Fig. 277), of a dark purplish brown, is covered above with rough wart-like tubercles, forming a prominent circle around the eyes. One of the species of the genus gives out a strong smell of musk.

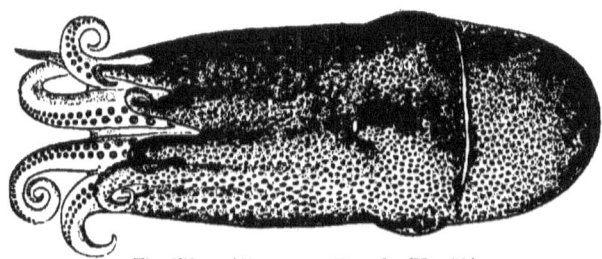

Fig. 278. — Alloposus mollis. ⅘. (Verrill.)

Another characteristic species is *Alloposus mollis* (Fig. 278), having a thick, soft, smooth body, and arms united by a web nearly to their extremity.

Along the Atlantic coast a number of cephalopods were dredged, many of them from considerable depths; among them we may mention Benthoteuthis. (Fig. 279.) They are mainly northern species, previously collected in shallower waters by the United States Fish Commission.

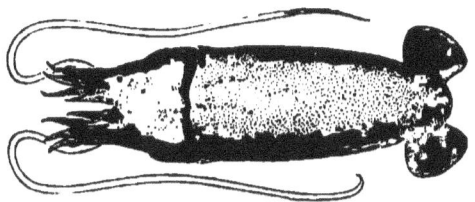

Fig. 279. — Benthoteuthis. ¼. (Verrill.)

But by far the most interesting of the cephalopods is a Spirula (Fig. 280) in excellent condition, dredged off Grenada in the

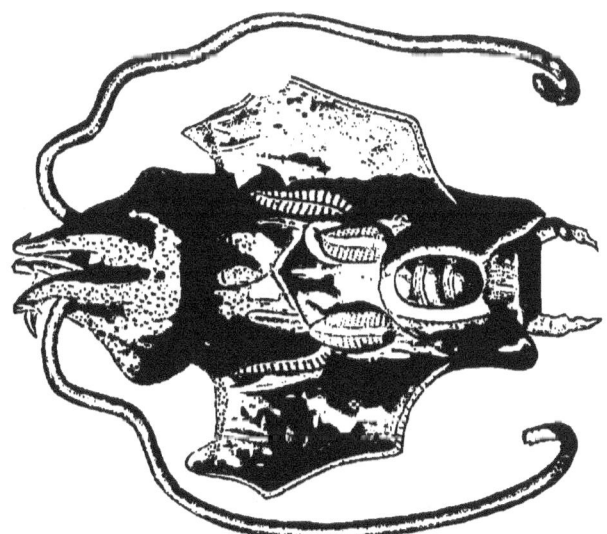

Fig. 280. — Spirula. ¹⁰⁄₇. (Huxley.)

Caribbean by the "Blake" from a depth of 950 fathoms. From the condition of the chromatophores of the body, it evidently lives with its posterior extremity buried to a certain extent in the mud. The "Challenger" collected a specimen from 360 fathoms, off the Banda Islands. Cephalopods have been collected by the Fish Commission off Martha's Vineyard from a depth of over 1,000 fathoms.

The giant squids (Fig. 281) of the North Atlantic (Architeuthis), occasionally thrown up on the shores of Newfoundland, attain an immense size, the arms measuring fully forty feet in length. They probably live in the regions where food is most abundant, upon the slopes, near the boundary of the continental plateau. It will be some time before we are able, with our present appliances, to capture such monsters from the depths at which they live. The Belemnites, so characteristic of some of the tertiary deposits, have not as yet been dredged.

GASTEROPODS AND LAMELLIBRANCHS.[1]

The Mollusca obtained by the "Blake" are notable in several respects. We may refer to the absence or rarity of very minute forms, which are only accidentally preserved in the contents of a trawl net, even from comparatively shallow water. It is hardly to be expected that, in the long washing which the contents of a trawl undergo while hauled in from deep water, anything small enough to go through the finest meshes of the bottom net should be retained. Yet large shells appear to be rare in the great depths, and are usually so fragile that their destruction or fracture is almost inevitable. Deep-sea dredging has thus afforded few specimens of even moderately large size, judged by the standard of shallow-water or littoral shells. Among naked mollusks several species of unusual size have been found by different expeditions. One as large as an orange, discovered by the "Challenger," was named by Dr. Bergh *Bathydoris abyssorum*. It is perhaps the largest nudibranch known; it has a transparent and gelatinous consistency, and with neither eyes nor otocysts it must have led a remarkably sluggish existence, blind and deaf as it was.

Abyssal mollusks are probably less active and energetic than their congeners of the shores. This is indicated by the looseness of the tissues, less favorable to prompt and violent action than a more compact muscular system would be. The tena-

[1] Mr. Dall has kindly prepared for me the account of the Gasteropods and Lamellibranchs, and supervised the drawing of the figures.

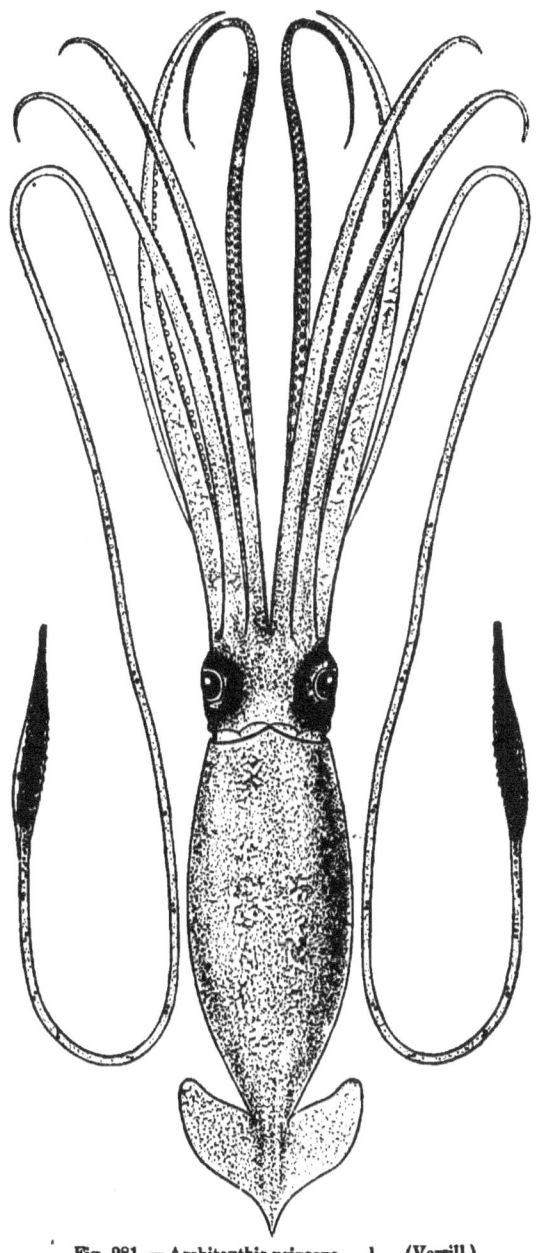

Fig. 281.— Architeuthis princeps. $\frac{1}{15}$. (Verrill.)

cious character of the mud forming the ocean floor would also tend to make motion through it slow and difficult. The delicacy of the shells, their extreme fragility and tenuity often reminding one of the delicate dwellings of some of the tropical land snails, would unfit them for constant friction and collision, either from the motions of the animal itself or of the waters in which it lives. Swimming mollusks, such as the squids and cuttle-fishes, make an exception; but the deep-sea representatives of these groups are far softer and less muscular than their shallow-water allies.

The colors of the abyssal shells are almost always faint, though often pretty. The iridescence or pearly character of the shell is in many groups of peculiar brilliancy and beauty, and it seems as if the texture of the non-iridescent shells in the abyssal species gave out a sort of sheen which is wanting in their shallow-water allies.

We do not find in the deep-sea species those sturdy knobs and stout varices which ornament the turbinellas and conchs of shallow water, and have made the great group of rock-purples, or Murices, so attractive to collectors; nevertheless many abyssal shells have an exquisite and rich sculpture, and their ornamentation is wonderfully delicate. There seems to be an especial tendency to strings of bead-like knobs, revolving striæ and threads, and delicate transverse waves. Many of the deep-sea forms, selected from all sorts of groups indifferently, have a row of knobs or pustules following the line of the suture and immediately in front of it. Their surface is also frequently etched with a sort of shagreen pattern, varied in detail and hardly perceptible except by a microscope, but extremely pretty. In some the entire surface is profusely adorned with arborescent prickles; in others, it is covered with the most delicate shelly blisters, systematically arranged, which perish with a touch.

Deep-sea mollusks may be understood to include all those living on the continental shelf, and in the abysses at depths where algæ do not flourish, the limit depending somewhat on the locality. Those living only above form the littoral fauna, which, roughly speaking, may be said to reach from the shores to about one hundred fathoms in depth. With them are often mixed deep-water forms, which extend their range to shallow water without however being characteristic of it.

As in other groups, the limits of many species of mollusks are more sharply defined on the side of cold than on that of heat. The difference between 45° and 40° F. may absolutely check the distribution of a species which would find no inconvenience in a rise of temperature from 45° to 80°. As has been observed in fishes, this limit is probably connected with the temperature necessary for development of the young, rather than with the resisting powers of the adult.

It would seem as if the conditions existing on the floor of the deeper parts of the ocean offered attractions for only a limited variety of forms. The bottom is generally composed of extremely fine impalpable mud, and in many portions of the abyssal area offers no stones or other prominences as points of attachment for sedentary mollusks. It is not quite destitute of such irregularities, however, and all are utilized by the abyssal population. In the absence of stones, most unusual selections are made. The chitinous tubes of hydroids and the irregular leathery dwellings of tubicolous annelids are occupied, after their original owners are dead or dispossessed, by diverse little limpets. The long spines of the abyssal sea-urchins offer a welcome perch for species of Cadulus, which, when they grow too large to find a satisfactory foothold, secrete a shelly pedestal which serves them for life.

A bivalve, *Modiolaria polita*, related to the ordinary mussel of northern seas, spins a sort of nest of stout byssal threads, in which it is completely concealed, and which protects in its meshes not only the young fry of the maker, but various little commensal mollusks of all orders. Only a small number of mollusks live as commensals. Species of Stylifer, a small gasteropod, live associated with star-fishes, sea-urchins, and other echinoderms. Dr. Stimpson discovered another living within an annelid; and they are often found imbedded in branches of corals, of which they have become a part as it were.

Those mollusks which live on algæ and other vegetable matters are almost absolutely wanting in the depths of the sea, where vegetation, except as a sediment from near the surface, does not exist, so that the flesh-eating mollusks of the deep, when within reach of pelagic food, or of the carcasses of dead fishes and other decaying organic matter, are not obliged to prey upon each other to the same extent as do the shallow-water forms. The latter take part in a fierce struggle for existence amidst the vicissitudes of tidal and storm waves, variation in elevation of land, and a vastly denser population of all sorts. Comparatively few of the shells dredged from deep water show the fractures and injuries so common in shells from littoral dredgings, or the drill-holes made by the so-called lingual ribbons, a terrible boring weapon of enemies of their own kind. Most of the enemies of deep-water mollusks are blind, or at any rate can have little power of vision for objects not luminous. The absence of violent motion in deep water removes any mechanical effects of that medium from the category of modifying influences upon the animal. Thus it is evident that the factors affecting the restriction of tendencies to variation in the form, color, and sculpture of littoral species are nearly eliminated in the abyssal regions; so that we may expect in the deep sea a very wide range of variation in form and sculpture within the specific limits of the "flexible" species, and an almost complete uniformity over very wide areas of the forms which we may consider as "inflexible" species.

Many of the gasteropods must lead a more or less roving life in search of their prey; others, like Dentalium, live buried in ooze. A great number of the mollusks are blind. The lamellibranchs live either buried in the ooze, or on the surface of harder bottoms anchored by the byssus. Most of them are stationary, though, judging from analogy with some of the shallow-water genera, they may be capable of considerable change of locality.

Those mollusks which subsist upon other animals, with a hard covering, so that they have to bore or break their way to their food, are much less numerous in the deep sea than those which feed upon soft tissues, or kill their living prey by bites with poisonous fangs. The latter, the Pleurotomidæ, outnumber any other group of mollusks in the abyssal fauna; they are characterized by a notch near the junction of the outer margin of the aperture with the outside of the preceding whorl. This notch permits the refuse matters discharged from the anal opening to escape outside of the shell without fouling the water which is used by the gills in respiration. These mollusks are found at all depths, are animal feeders, and some of them are provided with barbed hollow teeth, having a duct to which a gland supplies a poisonous substance; such an apparatus is even more fully and generally developed in the related group of Conidæ, few of which reach any great depth.

Among those Pleurotomidæ which would attract especial attention is the exquisite *Pleurotoma (Ancistrosyrinx) elegans* (Fig. 282), one of the most beautiful gems of the sea. It grows to an inch and a half in length, and is of a light straw color; the posterior surface of the whorls is concave and carinated, the carinæ being delicately fringed with sharp triangular points; it has a deep notch, which in perfect specimens has a raised margin. This species descends to eight hundred fathoms, and has been found alive at Barbados in seventy-three fathoms. Its fossil allies extend as far back as the eocene.

Fig. 282. — Pleurotoma (Ancistrosyrinx) elegans. ⅔.

Fig. 283. — Pleurotoma subgrundifera. About ¾.

Pleurotoma subgrundifera Dall (Fig. 283) is a form which, instead of having the margin turned toward the tip of the spire, has the sharp keel bent in the opposite direction toward the canal, like the edge of an umbrella. Another pretty species, dredged in deep water both by the "Blake" and the "Challenger," is *Pleurotoma Blakeana;* and still another, short and stout, with delicate reticulate sculpture, has also been obtained by the Fish Commission, the *P. curta* of Prof. Verrill. Both these resemble in shape the Belas of the arctic seas. A very elegant and widely distributed little shell is the *P. limacina*, polished, smooth, with a beaded garland at the suture; it is extremely thin, with peculiar flexuous growth lines and no operculum. The variety in this group seems endless, and in number of species it is likely to rival even some of the great groups of land shells.

The groups of less specialized character, such as the tusk-shells (Dentalium), are rather abundant in species, more so than those which intervene between them and the highly specialized Pleurotomidæ; but our knowledge of the deep-sea mollusks is yet too imperfect to afford any important generalizations on this score. So far as determined, the groups systematically lowest in the scale, like the Chitonidæ or mail-shells, are rare in deep water, yet the deep-sea representatives of this family belong to the more archaic sections of their class. The tusk-shells are curved tubes, almost all white or delicately tinted, and varying chiefly in curvature, calibre, and superficial sculpture or color. The most remarkable of these, among the slender species, is *Dentalium perlongum* (Fig. 284), polished, white, nearly smooth, and attaining a greater relative length than any other species, over four inches, with a diameter of an eighth of an inch at one end, and half as much or less at the other. It reaches the greatest depths dredged by the "Blake" (over 2,000 fathoms), and has not appeared in shallow water. There are many other species, but it is only necessary to mention one peculiar group of the family, the genus Cadulus, containing numerous species, all of which are small, polished, pellucid shells. They expand their little tubes to a sort of bulb, more or less prominent, which diminishes before they are completed, so that the calibre of the aperture is smaller in the adult than in the young; while in the true Dentalium the diameter gradually increases with age. The Caduli are quite characteristic of the deeper waters of the sea.

Fig. 284. Dentalium perlongum. †.

Another group also largely represented in the abyssal region is that of the Trochidæ. These are among the most beautiful of spiral shells, often brilliantly colored, profusely sculptured, and very pearly. The shallow-water forms may subsist on stony algæ or other plants, but the majority are flesh-eaters, or feed upon the corallines and foraminifers, parts of whose shells are found in their stomachs.

While not so brilliantly colored, the deep-water Trochidæ are unsurpassed in beauty by their shallow-water allies. They gain in delicacy and iridescence what they lose in depth of tint. One of the handsomest forms is *Calliostoma Bairdii* Verrill, whose pale, depressed, and more delicate southern variety, *C. psyche*, was first dredged by Pourtalès. It is, like many other species of similar range, tinted with pink and straw-color, while farther north it assumes brown and red livery. Even more delicate and peculiar in the concave outline of its granular spire and polished base is *Calliostoma aurora* (Fig. 285), of which only a single specimen is known, — a genus most characteristic of Western America. It seems as if differences of temperature and food were indicated in very similar ways between northern and tropical animals, whether they live in the deep sea or inhabit the land.

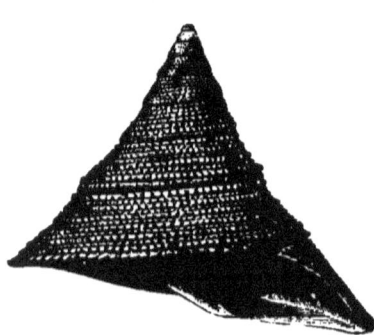

Fig. 285. — Calliostoma aurora.

A real treasure of the sea is *Gaza superba* (Fig. 286), one of the most beautiful and widely distributed abyssal shells. Were it not for its lovely iridescent pearly sheen, it might be taken, on a casual examination, for one of our large straw-colored land snails.

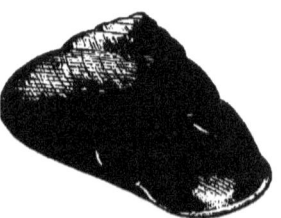

Fig. 286. — Gaza superba.

Other characteristic species, widely distributed, are *Margarita ægleës* and *Leptothyra induta* (Fig. 287) of Watson, small white shells from deep water, named from examples collected by the "Challenger," and especially illustrating the luxury in variation which has already been referred to, and which has led in the case of the former to the application of several specific names. The depth in which these have been found varies from 125 to over 1,000 fathoms.

Pleurotomaria is one of the most remarkable forms dredged in the continental region. Four recent species of the genus are known. Its history dates back to the earliest fossiliferous rocks of the cambrian, and to the dredgings of the "Hassler" and the "Blake" are due the only knowledge yet acquired of its soft parts. Two species are found in the

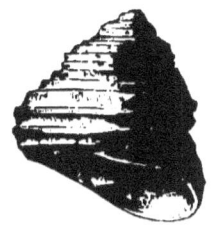

Fig. 287. — Leptothyra induta. ⅓.

West Indies, of which the finest is *P. Adansoniana* (Fig. 288), from about 200 fathoms. The shell is four inches in diameter, richly pearly within, and ornamented with elegant red and brown colors externally. The anal notch in this species extends nearly half the length of the last whorl. A second species, less brilliant and with a shorter notch, is *P. Quoyana* (Fig. 289), also obtained living by the "Blake."

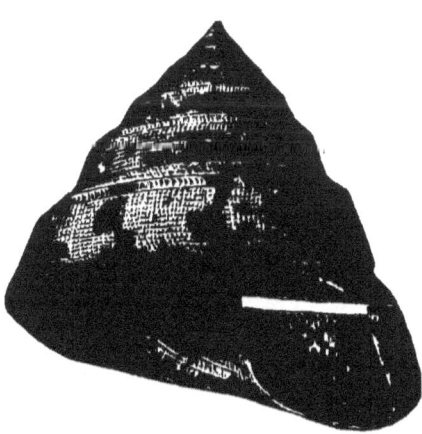

Fig. 288. — Pleurotomaria Adansoniana. ⅔.

Among other univalves, the Marginellidæ are represented by such species as *Marginella succinea* Conrad, extending from shallow water to several hundred fathoms, and *M. Watsoni* (Fig. 290), characteristic of great depths. The Ringiculidæ, of which many species are known fossil, are illustrated by *R. leptocheila* (Fig. 291), described first by Brugnone from the Mediterranean, and afterward from deep water

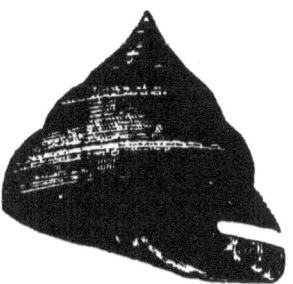

Fig. 289. — Pleurotomaria Quoyana. ⅓.

Fig. 200. — Marginella Watsoni. ⁸⁄₁.

Fig. 201. — Ringicula leptocheila. ⁴⁄₁.

in the Bay of Biscay and on the coasts of America. *Cancellaria Smithii* (Fig. 292), an elegant new species of a comparatively rare group; *Mitra Swainsoni* (Fig. 293) of Broderip, from the deep water of the West Indies, first described from Chilian waters; and *Typhis longicornis* (Fig. 294), a pretty flesh-colored deep-water species, — may be cited as examples of other groups, the last being particularly remarkable from the length of its spines, which could only exist in the shell of an animal surrounded by a soft bottom and living in perfectly calm water.

Fig. 292. Cancellaria Smithii. ⁷⁄₁.

Fig. 293. — Mitra Swainsoni. ⁴⁄₁.

Fig. 294. — Typhis longicornis. ²⁄₁.

Another illustration of the fragile and delicate forms living in the abysses is *Triforis longissimus* (Fig. 295), only thir-

Fig. 295. — Triforis longissimus. ⅔.

teen hundredths of an inch in diameter, with a column of twenty or thirty whorls, reaching an inch to an inch and a half in length; the perfect shell must have over forty turns, but it is always decapitated. *Siliquaria modesta* (Fig. 296),

 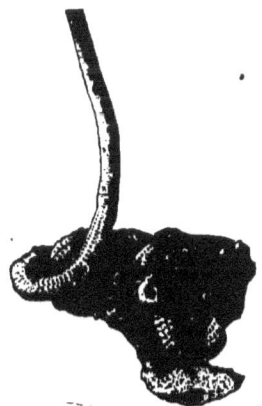

Fig. 296. — Siliquaria modesta. 1/1.5.
Fig. 297. — Vermetus erectus. 1/1.5.

one of the irregular gasteropods, with a slit like a Pleurotomaria, so frail as almost to perish with a touch, lives in the soft mud of the abysses, while the stouter *Vermetus erectus* (Fig. 297) finds a foothold on dead corals and shells. The species of this genus are comparatively shallow-water animals.

The majority of the bivalves are characterized by great delicacy of shell and sculpture. In the deep-water representatives of the family of scallops, the constituent prisms are often large enough to be seen with the naked eye, and the shell is strengthened within by slight riblets radiating from the hinge. *Pecten*

Dalli (Fig. 298), of E. A. Smith, frequently dredged by the "Blake," grows to a considerable size, but is as thin as mica and nearly as transparent; *P. phrygium* Dall (Fig. 299) is re-

Fig. 299. — Pecten phrygium. ¹⁄₁.

Fig. 298. — Pecten (Amusium) Dalli. ¼.

lated to miocene species, and has a very complicated sculpture. *Cetoconcha bulla* (Figs. 300, 301) and *C. elongata* Dall (Fig.

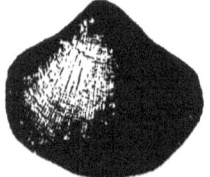

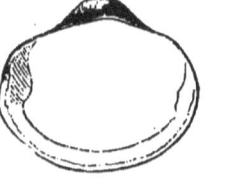

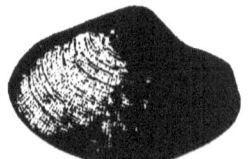

Fig. 300. Fig. 301.
Cetoconcha bulla. ⅔.

Fig. 302. — Cetoconcha elongata. ¹⁄₁.

302), two species of a singular new genus, are almost as unstable in their framework as a drop of water. Nuculæ are abun-

Fig. 303. — Tindaria cytherea. 1⅕.

Fig. 304. — Cardium peramabilis. ⅔.

dant. One of them, *Tindaria cytherea* (Fig. 303), the finest and largest of its genus, white, with a golden epidermis, is peculiar in its shape, which resembles that of a small member of the Veneridæ. A delicately sculptured Cardium, sometimes painted with bright touches of yellow and scarlet, *Cardium peramabilis* (Fig. 304), the most lovely species of the genus from deep water, shares with the little *Pecten (Amusium) Pourtalesianum* Dall the distinction of bright tints where pallor is the rule. The shell is white, but the spines covering it are orange or crimson. A common and characteristic deep-water form is *Limopsis aurita* Brocchi, well known as a tertiary fossil in Europe. A small brown Astarte is almost ubiquitous, ranging in depth from 13 to over 1,600 fathoms, and in locality from the tropics to New England. The northern specimens attain many times the size of those from the Antilles. A highly polished rich golden brown Modiola, *M. polita* V. & S. (Fig. 305), allied to our common mussel, attains a large size in great depths on both sides of the Atlantic. But its shell is very thin; it spins a large nest of byssal threads, resembling a handful of cotton waste thoroughly drenched with the finest mud, so worthless in appearance that only a biologist would suspect the treasure hidden within.

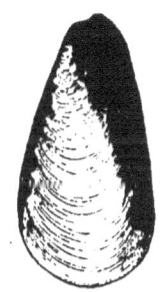

Fig. 305. — Modiola polita. ⅔.

The Cetoconcha above mentioned are characterized by gills reduced to a mere interrupted line of low lamellæ on the ventral surface; they are related to Poromya, which has ordinary gills. But there is another group, abundant in deep water, called Cuspidaria, still more remarkable in having apparently no gills at all; their shells are provided with a long slender rostrum, like a handle, as shown in *C. microrhina* Dall (Figs. 306, 307), dredged from continental depths. A striking group, from the beauty of form and sculpture exhibited by its species, is Verticordia, the

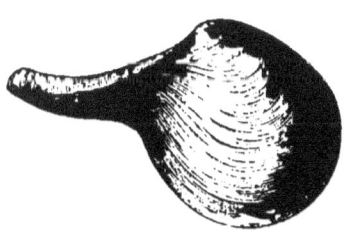

Fig. 306. — Cuspidaria microrhina. ⅔.

largest known species being one of the "Blake" treasures, *V. elegantissima* Dall (Fig. 308), a brilliant pearly shell; one of the smallest is *V. perversa* Dall (Fig. 309), which has the larger bulge in front of the hinge, contrary to the usual rule.

Fig. 307. — Cuspidaria microrhina. ½.

A lovely new group related to Thracia and Anatina is repre-

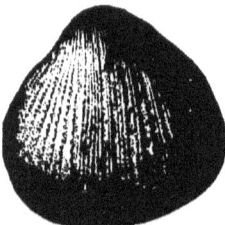
Fig. 308. — Verticordia elegantissima. ½.

Fig. 309. Verticordia perversa. ⅔.

Fig. 310. — Bushia elegans. ⅔.

sented in deep water by a single species, which has been named *Bushia elegans*. (Fig. 310.)

We may also mention, as evidently a deep-water group, the shells of the subgenus Meiocardia, related to *Isocardia cor* of Europe. These are remarkable for the way in which the tips of the valves are twisted and turned away from each other. They are common tertiary fossils; but only a few living species are known, and, excepting *Isocardia cor*, these are tropical.

Fig. 311. — Meiocardia Agassizii. 1;0.

The dredgings of the "Blake" and the "Albatross" have revealed a new Meiocardia in the Antilles, the others being all Oriental, and this has been named *M. Agassizii*. (Fig. 311.)

A new group, differing from Isocardia and Meiocardia in having no lateral teeth, is Vesicomya, previously unknown from American waters, the largest known species of which is a form now named *V. venusta* (Fig. 312), from Antillean specimens. A much smaller species, named *V. pilula*, is reported by the

"Challenger" from the deepest water in which any bivalve has yet been found living.

There are almost innumerable illustrations of beauty, adaptation, or unusual characteristics which might be cited, but to those unacquainted with the objects themselves such an enumeration would be tedious. The enthusiastic student and collector alone can find pleasure in what would seem to most people a dry combination of a lexicon and a catalogue.

Fig. 312. — Vesicomya venusta. $\frac{1.8}{1}$.

BRACHIOPODS.

Until quite lately brachiopods were rarities in collections; but since the days of dredging expeditions we know that they are very numerous at favorable localities on rocky or stony bottoms. They do not seem to penetrate very great depths, naturally finding no point of attachment in the soft ooze of the deep waters, and but few species are thus far known to extend beyond 600 fathoms. The largest known species have been dredged from the abyssal region, and young specimens are frequently found attached to the older ones. None of the deep-water species have the brilliant coloring characteristic of the common littoral species belonging to the genus Lingula. The principal differences upon which their classification is based are those of the so-called loop, the calcified support of the brachia, and the structural details of the valves.

The recent brachiopods are specially interesting as representatives of a group which attained an extraordinary development in very early ages, and has been represented in all formations. They have a most extensive geographical distribution, and a great bathymetrical range. They are found at all levels, from pools left by the tides to a depth of 3,000 fathoms. The number of living species is small compared to the hosts which flourished in the silurian, devonian, and carboniferous, from which time they have steadily diminished in number. Nearly 1,700 species occur in the silurian, but there are not more than 120 known from the seas of the present day. Their position in the animal kingdom is under discussion, though they have until lately been generally classed with the mollusks. As the brachiopods date back to the cambrian, it is natural that we should find it difficult strictly to define their affinities with recent types, since with very slight modifications they have persisted from remote antiquity to the present day, during all the intervening conditions of existence.

Like the lamellibranchs, they are provided with two valves. These, however, as well as the soft parts, are bilaterally symmetrical in relation to the longitudinal axis of the shell.

The most common species we collected, *Terebratula cubensis* (Figs. 313, 314), was discovered by Pourtalès, in from 100 to

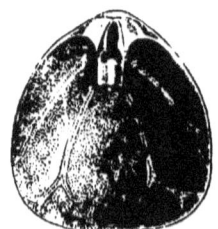

Fig. 313. $\tfrac{1}{1}$.

Fig. 314. $\tfrac{1\cdot 25}{1}$.

Terebratula cubensis. (Davidson.)

270 fathoms, in rocky ground off Havana and from the east end of the Florida Reef. It attaches itself by a short and stout peduncle; the shell is globular, nearly white, translucent. Another most abundant species associated with the former is *Waldheimia floridana* (Figs. 315, 316), which is common on rocky

Fig. 315.   Fig. 316.
Waldheimia floridana. $\tfrac{1}{1}$.

bottoms between 100 and 200 fathoms. It is of a grayish or brownish white horn-color, and belongs to a group containing many living and fossil species. Much less common, but with a more solid test, is *Terebratulina Cailleti*. (Fig. 317.) This small species extends to a depth of nearly 500 fathoms. A most common Atlantic species, *T. caput-serpentis* (Fig. 318), is found along the eastern coast of the United States as far south

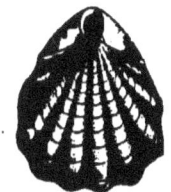
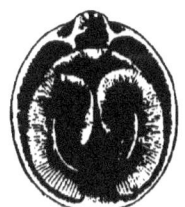

Fig. 317. — Terebratulina Cailleti. ⅔. (Davidson.) Fig. 318. — Terebratula caput-serpentis. ⅖. (Davidson.)

as Cape Cod. A species of Platydia, identical with the Mediterranean *P. anomioides* (Figs. 319, 320), has been dredged by the "Blake" in 237 fathoms. It represents the group of brachiopods with shells having loops and conspicuous perforations.

Fig. 319. Fig. 320.
Platydia anomioides. ⅔. (Davidson.)

A few specimens of Crania (Fig. 321), a genus not before obtained on the American coast, were dredged by Pourtalès off the Samboes and Sand Key, at depths ranging between 100 and 200 fathoms. Living specimens of *Discina atlantica* (Fig. 322) have been taken by the "Blake" and by the Fish Commission at the depth of over 2,000 fathoms. They are usually attached to concretions.

Fig. 321. Crania Pourtalesii. ²⁄₁. (Dall.) Fig. 322. — Discina atlantica. ⅔. (Verrill.)

The simple and compound ASCIDIANS are eminently littoral and shallow-water types, and but few of them extend to any great depth. Neither the "Blake" nor the "Challenger" collected any very remarkable abyssal types, and the species were either closely allied to or identical with well-known genera.

BRYOZOA.

In the study of no group is abundant material more necessary than in that of the bryozoans. In the majority of animals, we are accustomed to look at dif-

Fig. 323. — Crisia denticulata. $\frac{2}{1}$.

Fig. 324. — Diastopora repens. $\frac{1}{1}$.

Fig. 325. — Farciminaria delicatissima. $\frac{1}{1}$. (Busk.)

ferences due to growth as transitory, and we define species from their full-grown stages; but in the bryozoans the differences of growth are persistent in the individuals of the colony, while they may propagate at very different stages of the colonial development. It

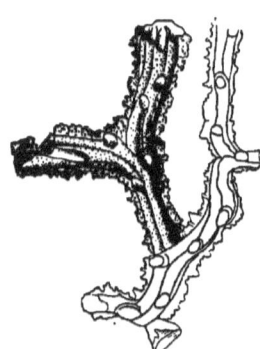

Fig. 323 a. — Crisia denticulata. Magnified. (Smitt.)

Fig. 324 a. — Diastopora repens. Magnified. (Smitt.)

thus becomes most difficult, without a full knowledge of the entire development, to characterize a species and assign it to its true family or genus. Among the bryozoa, more than three quarters

Fig. 325 a. — Farciminaria delicatissima. Magnified. (Busk.)

of the deep-water species belong to the section of the Cheilostomata, while the Ctenostomata have comparatively few representatives. Busk says that the shallower-water species appear to have the widest geographical distribution. That is apparently not the case with the species collected by the "Blake."

According to Professor Smitt's Reports we may mention among the "Blake" Bryozoa the cosmopolitan *Crisia eburnea*, the form known as *C. denticulata* (Figs. 323, 323 a), and, from 306 fathoms, the Scandinavian *Diastopora repens* (Figs. 324, 324 a), a well-known ramified form creeping on *Terebratula cubensis*. This species is also characteristic of the crag, and perhaps identical with a cretaceous form. It seems as if the species of this group assumed a somewhat more elongate and simpler form in proportion to their bathymetrical range. Busk, from an examination of the extensive collection of the "Challenger," considers the species of Farciminaria (Figs. 325, 325 a) as the most characteristic of the abyssal bryozoans, the preëminent forms of the delicate and flexible types inhabiting the tranquil depths of the ocean.

Membranipora canariensis (Fig. 326), a widely spread spe-

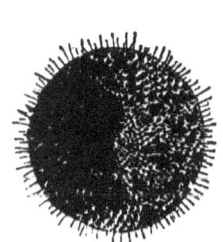

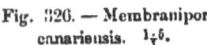

Fig. 326. — Membranipora canariensis. $\frac{1.6}{1}$.

Fig. 327. — Cellularia cervicornis. $\frac{2}{1}$.

Fig. 327 a. — Cellularia cervicornis. Magnified. (Smitt.)

cies, found in both hemispheres, and common in the tertiaries[1]

---

[1] There are among the Florida and West Indian bryozoans no less than sixteen species identical with those of the tertiary period, and about five either the same or closely allied to cretaceous types.

of Sicily and England, is abundant off Florida to a depth of over 120 fathoms. It generally takes the shape of a hollow cone. Among the Cellulariæ, *Cellularia cervicornis* (Figs. 327, 327 a) and *Caberea retiformis* (Fig. 328) are interesting repre-

Fig. 328. — Caberea retiformis. Magnified. (Smitt.)

Fig. 329. — Vincularia abyssicola. ⅔.

sentatives, the last closely allied to a typical Australian species. Other species of this group are similarly allied to Australian types. *Vincularia abyssicola* (Fig. 329), from 450 fathoms, is a most variable species, likely to be placed even in distant families

Fig 330. — Eschariporu stellata. ⅔.

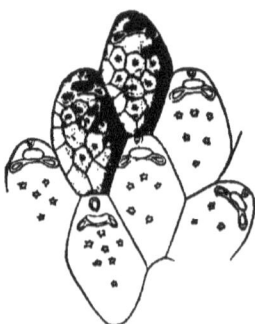

Fig. 330 a. — Escharipora stellata. Magnified. (Smitt.)

at different periods of its growth, while either in the creeping or in the erect stage. *Escharipora stellata* (Figs. 330, 330 a)

CHARACTERISTIC DEEP-SEA TYPES. — BRYOZOA. 81

is one of the most common of the West Indian bryozoans inside the two-hundred-fathom line, but extending to nearly 500 fath-

Fig. 331. — Tessadroma boreale. ⅓.

Fig. 331 a. — Tessadroma boreale. Magnified. (Smitt.)

oms. Equally common is *Tessadroma boreale* (Figs. 331, 331 a), a species not infrequent on the east side of the Atlantic from Spitzbergen to the Azores.

A very common incrusting type found growing on shells and corals is *Hippothoa biaperta* (Figs. 332, 332 a), which goes back to the tertiary.

A species of Cellepora is very abundant

Fig. 332 a. — Hippothoa biaperta. Greatly magnified. (Smitt.)

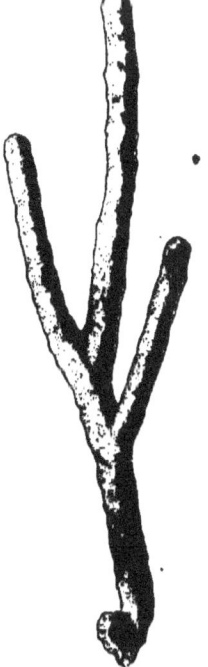

Fig. 332. — Hippothoa biaperta. ⅓.

in depths ranging from 15 to nearly 300 fathoms, *C. margaritacea.* (Figs. 333, 333 *a.*)

Fig. 333. — Cellepora margaritacea. ²⁄₁.

Fig. 333 *a.* — Cellepora margaritacea. Magnified. (Smitt.)

Among the Bryozoa often found in large communities, forming lawns of delicate limestone plants, may be specially men-

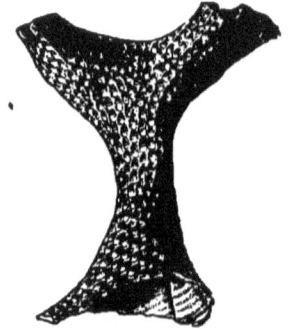

Fig. 334. — Biflustra macrodon. ⅔.

Fig. 335. — Porina subsulcata. ⅔.

tioned *Biflustra macrodon* (Fig. 334), *Porina subsulcata* (Fig.

Fig. 336. — Retepora reticulata. ½.

335), and *Retepora reticulata.* (Fig. 336.)

A supposed Favosites (Fig. 337), mentioned in the prelimi-

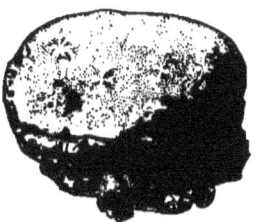

Fig. 337.—Heteropora. ¼.

nary accounts of the results of the "Blake" expeditions, is probably a bryozoan genus growing in the shape of a mushroom and allied to Heteropora.

XIX.

CHARACTERISTIC DEEP-SEA TYPES.—ECHINODERMS.

HOLOTHURIANS.

THE order of Apoda among holothurians has neither pedicels nor suckers, while the Pedata have a highly developed ambulacral system and a well-defined dorsal and ventral surface, with pedicels scattered over the whole body. The large lobes of the Elasipoda (the new order of deep-sea holothurians established by Dr. Théel) are perhaps tactile. The ventral surface of the Elasipoda is intended for locomotion, and, as suggested by Dr. Théel, they probably move along the bottom with the actinostome wide open, constantly filling their alimentary canal with the ooze stirred up by the tentacles of the mouth. The calcareous deposits resemble those of the larval holothurians, and they possess other features showing them to be an embryonic type. The auditory capsules are often present in great numbers.

The Elasipoda are strictly abyssal types, no member of the group having been dredged in less than 50 fathoms, and that only in the Arctic Ocean, where, as we know, deep-sea types are found in comparatively shallow water. Of the large number of "Challenger" species, only five are found within the 500-fathom line, as many more inside the 1,000-fathom line, and the others all below that limit. At the localities where the "Blake" was fortunate enough to find Elasipoda, they occurred in large numbers, and, judging from the contents of the trawl, they apparently live in communities including several species, and prefer soft ooze. The experience of the "Challenger" and of the Fish Commission was a similar one. The "Challenger" obtained on one occasion no less than ten species associated together.

Owing to the absence of fossil holothurians we are unable, as in the case of other echinoderms, to trace the groups from which this peculiar deep-sea order of Elasipoda has been derived. While during earlier geological periods the holothurians undoubtedly made their way by gradual migration from the shore into deep water, their shallow-water progenitors have left us no

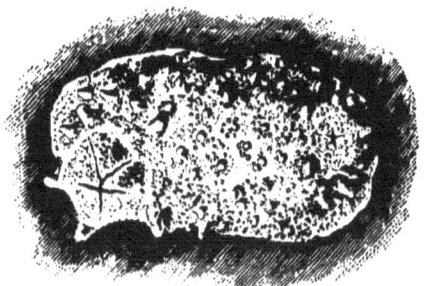

Fig. 338. Psolus tuberculosus. ⅔. (Théel.)

trace of their existence. The whole tribe of Elasipoda, which stands out apparently isolated from the other orders of holo-

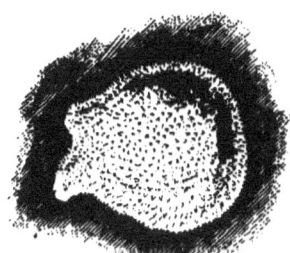

Fig. 339. — Echinocucumis typica. ⅔. (Théel.)

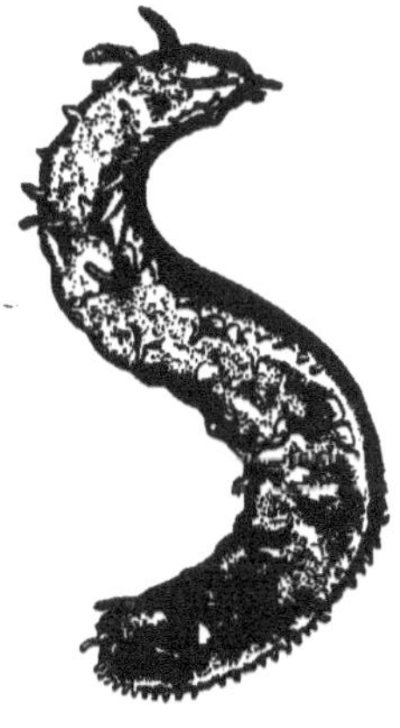

thurians, is found associated with such genera as Psolus (Fig. 338), Echinocucumis (Fig. 339), Stichopus (Fig. 340), Trochostoma[1] (Fig. 341), and Caudina, all of which have representatives in deep water, and some even in very deep water.

Fig. 340. — Stichopus natans. ⅓. (Koren & Danielssen.)

[1] *Trochostoma arcticum* is of a greenish violet color; the tentacles are much lighter; and the skin is comparatively tough.

Of the Elpididæ proper, the family of the order first described by Théel, no representative was dredged by the "Blake"; but

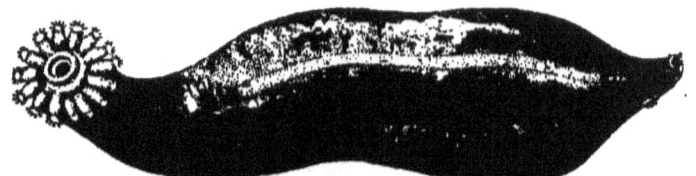

Fig. 341. — Trochostoma arcticum. ⅔. (Koren & Danielssen.)

the Deimatidæ and Psychropotidæ (Fig. 342) are both in the "Blake" collections. In the last-named family there extends

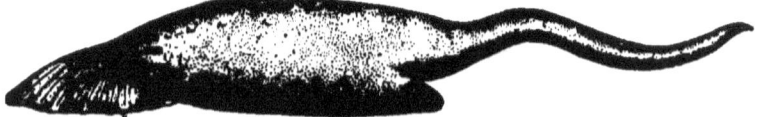

Fig. 342. — Psychropotes longicauda ¼. (Théel.)

round the body a more or less distinct margin edged by numerous lateral pedicels of small size, while in the Deimatidæ (Fig. 343) these are large and few in number.

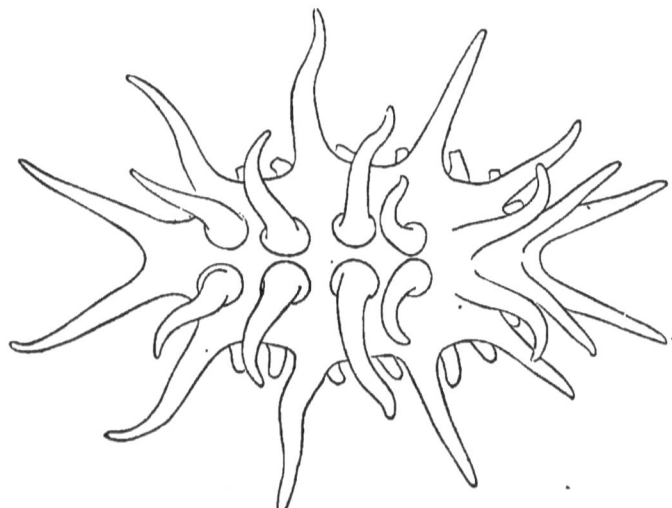

Fig. 343. — Deima Blakei. ⅖. (Théel.)

CHARACTERISTIC DEEP-SEA TYPES. — HOLOTHURIANS. 87

Huge species of Elasipoda were found in great numbers at several stations beyond 1,000 fathoms. The same species were also dredged by the U. S. Fish Commission, and the drawings here given of the gigantic Benthodytes and Euphronides I owe to the kindness of the Fish Commissioner, Professor Baird. Benthodytes (Fig. 344) is flat below, convex above, of a trans-

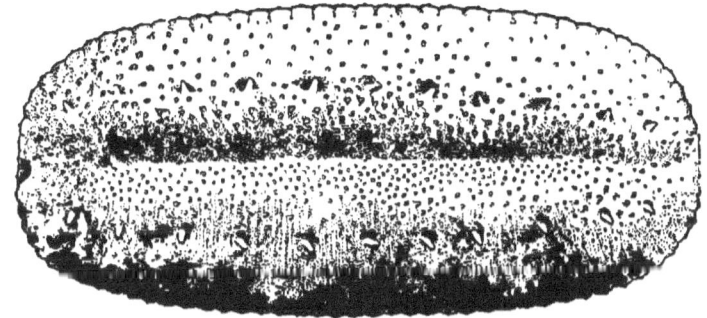

Fig. 344. — Benthodytes gigantea. ⅔. (U. S. F. C.)

lucent appearance, but of considerable consistency when fresh. It was most difficult to preserve these huge holothurians in alcohol, and the specimens sent to Dr. Théel were too imperfect for study. Euphronides (Fig. 345) resembles Benthodytes in

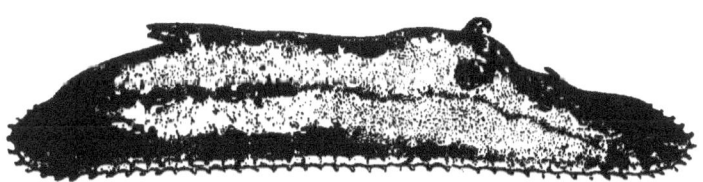

Fig. 345. — Euphronides cornuta. ⅔. (U. S. F. C.)

general outline, but in profile is high, slopes anteriorly and posteriorly, and has a lobed posterior appendage and a series of appendages placed in pairs on the bevel of the anterior extremity. It is of a reddish brown color.

Pælopatides (Fig. 346) and Ankyroderma[1] (Fig. 347) seem

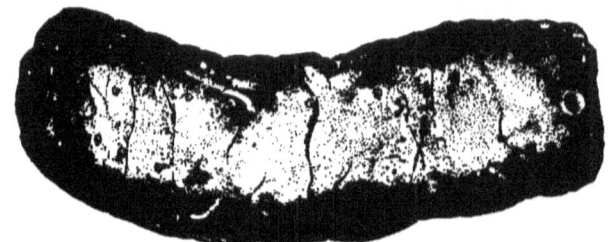

Fig. 346. — Pælopatides confundens. ⅔. (Théel.)

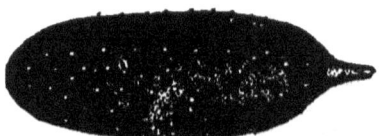

Fig. 347. — Ankyroderma affine. ⅔.
(Koren & Danielssen.)

to be the only typical truly deep-sea genera of the orders of Apoda and Pedata collected by the "Blake," not before found in the littoral regions, while the other deep-sea species belonging to genera found in shallow water are merely specifically distinct from the littoral forms, though undoubtedly, like other marine animals capable of living at extreme depths, they have become accustomed to their different conditions of existence most gradually, and those which live in deep water have acquired characters and habits somewhat distinct from those dwelling in the more littoral regions, but which a close study alone would reveal.

SEA-URCHINS.

One of the most common sea-urchins is *Dorocidaris papillata* (Fig. 348), a type having a very wide geographical distribution; it is found everywhere in the Atlantic, and has even been dredged in the Pacific; it came up in the dredge often to the exclusion of all other forms. It recalls a cretaceous type common both in Europe and America. As in all the Cidaridæ, the shape, proportions, and ornamentation of the spines vary greatly, and an exaggerated importance has frequently been assigned to char-

[1] *Ankyroderma affine* when alive is of a grayish color, the integument is thin, and the extremities of a lighter hue than the body.

acters derived from the study of a limited number of specimens, both in the fossil and recent species. In the seas of the Jura

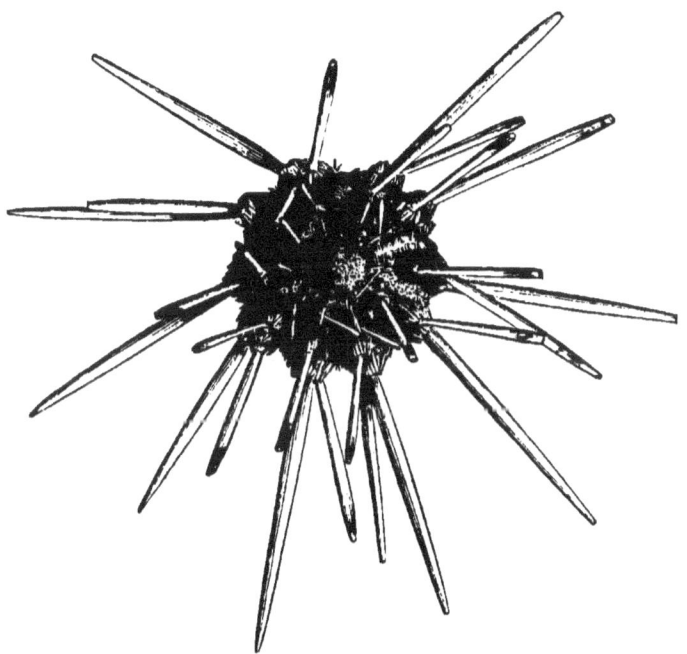

Fig. 348. — Dorocidaris papillata. ⅔.

and of the chalk the Cidaridæ must have been common types of sea-urchins. *Dorocidaris Blakei* (Fig. 349), obtained by the "Blake," is perhaps the most interesting of the recent Cidaridæ, from the variability of its spines. Before the "Blake" dredgings none were known among the recent species showing any great or striking variety in the form of the radioles. With the exception of some of the species of the genus Goniocidaris, the radioles are characterized by their uniformity, while among the fossils of the family the variation in shape and size of some of the jurassic and cretaceous species is quite remarkable. If the present species had been dredged without its two or three fan-shaped spines, it would have been unhesitatingly placed in the genus Dorocidaris. If the isolated huge fan-shaped radioles

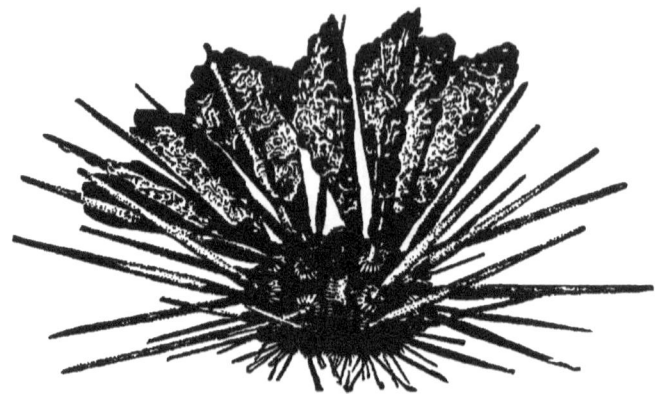

Fig. 349. — Dorocidaris Blakei. ⅔.

nearly identical in shape with those of the jurassic Rhabdocidaris had alone been collected, few palæontologists would have hesitated to refer them to that genus.

Another interesting type of deep-sea Cidaridæ allied to tertiary forms is Porocidaris (Fig. 350), which is characterized by the peculiar serrated spines found near the mouth.

We first dredged off Havana, and subsequently in all parts of the Caribbean, a fine species of Salenia (Fig. 351), a genus once very common in the jurassic and cretaceous seas. The first living species of the genus (Fig. 352) was dredged by Pourtalès

Fig. 352. — Salenia varispina. ¼.

off Double-headed Shot Key, in 315 fathoms. The "Blake" found it to be a characteristic species of the Caribbean abyssal fauna. This genus is characterized by the presence of a large

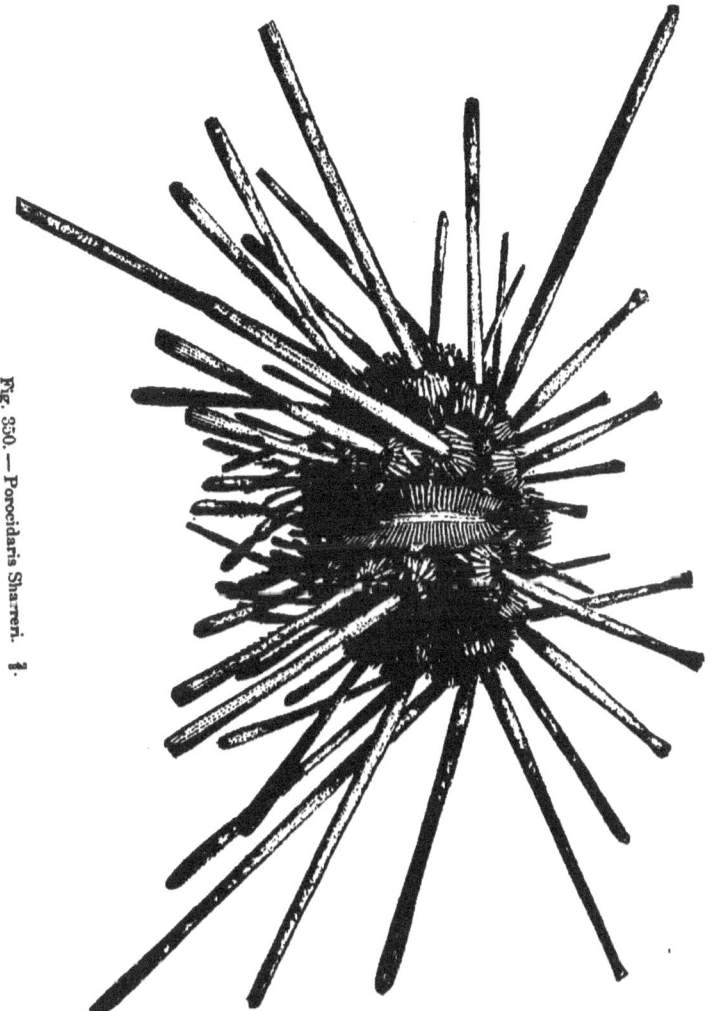

Fig. 350. — Porocidaris Sharreri. ⅔.

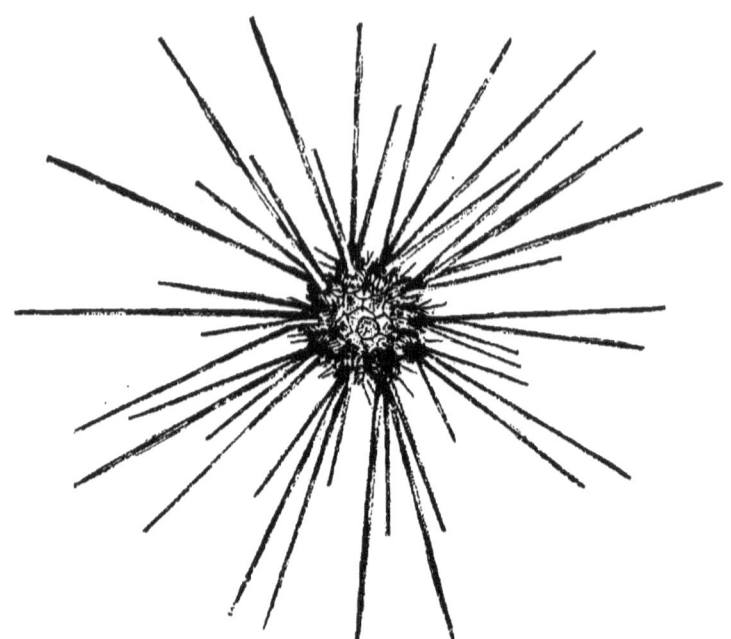

Fig. 351. — Salenia Pattersoni.

CHARACTERISTIC DEEP-SEA TYPES. — SEA-URCHINS. 91

suranal plate of an asymmetrical apical system (Fig. 353), combined with an arrangement of tubercles and of peculiar spines which connect it on the one side with the Cidaridæ, and on the

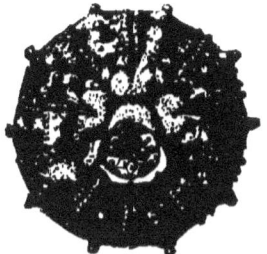

Fig. 353. — Salenia Pattersoni. ⅙.

other with the more recent types of sea-urchins. This asymmetry is an embryonic character of echinoderms, due to the spiral disposition of the plates of the embryo. Traces of this arrange-

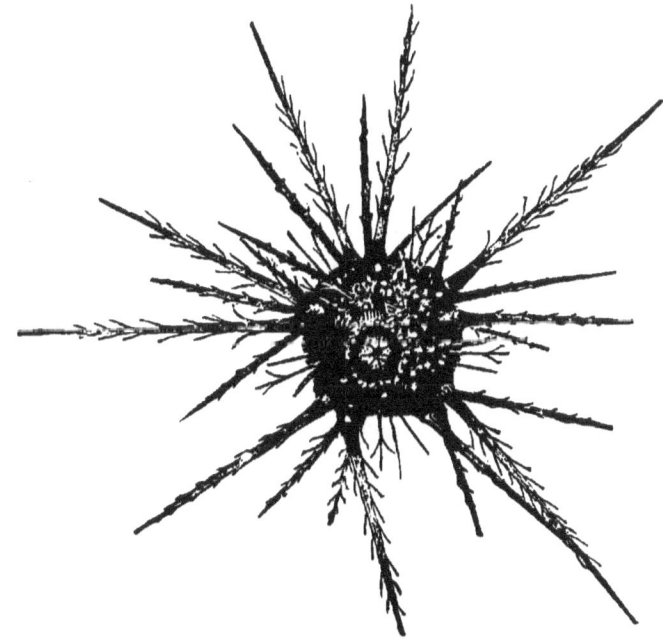

Fig. 354. — Salenia varispina. ⁹⁄₁.

ment are plainly to be seen in the unequal development in the size of the genital and ocular plates throughout the group of echini. Perhaps we may trace the differences in the development of the ambulacral and interambulacral zones in the echini to such a primitive differentiation. This embryonic feature runs back through the echinoid series of the earlier palæozoic times, and I am inclined to look upon the suranal plate of Salenia as recalling the crinoidal affinities of the sea-urchins, though it has not taken in the development of these the important part which it occupies in the starfishes and crinoids. The spiny primary radioles of the large specimens are formed from the gradual wearing of the delicate filaments (Fig. 354) of the corresponding spines in younger specimens.

As representatives of the sculptured echini so common during the tertiaries, and still prominent in the Indo-Pacific fauna, we find the small Temnechinus (Fig. 355) and Trigonocidaris. (Fig. 356.) The Arbaciadæ, a family of sea-urchins

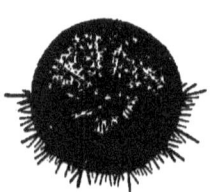

Fig. 355. — Temnechinus maculatus. $\frac{2}{1}$;^{5}.

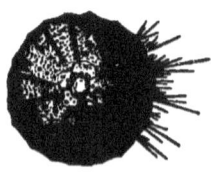

Fig. 356. — Trigonocidaris albida. $\frac{2}{1}$.

eminently characteristic of the American fauna, both Atlantic and Pacific, are represented in deep water by a highly sculptured genus, Podocidaris (Fig. 357), with primary spines recalling the embryonic ones of the littoral species. The large spines of these genera are used for locomotion, and for protection are tipped with a sort of shoe, which is constantly replaced as it wears. This shoe takes an immense development in Cœlopleurus (Fig. 358), and grows to three or five times the length of the spine itself. The primary spines are also curved, and when the urchin is in motion it is raised far above the surface, literally walking on stilts. The deep-water species must by means of their spines

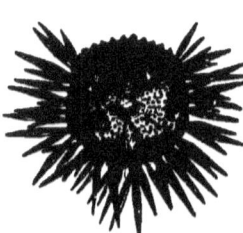

Fig. 357. — Podocidaris sculpta. $\frac{1}{1}$;^{5}.

CHARACTERISTIC DEEP-SEA TYPES. — SEA-URCHINS. 93

be capable of very rapid movements, if they at all correspond to those of their shallow-water allies.

As it is brought to the surface Cœlopleurus is most brilliantly colored, the test varying from a rich light chocolate in the interambulacra to the brilliant orange or yellow ambulacral areas.

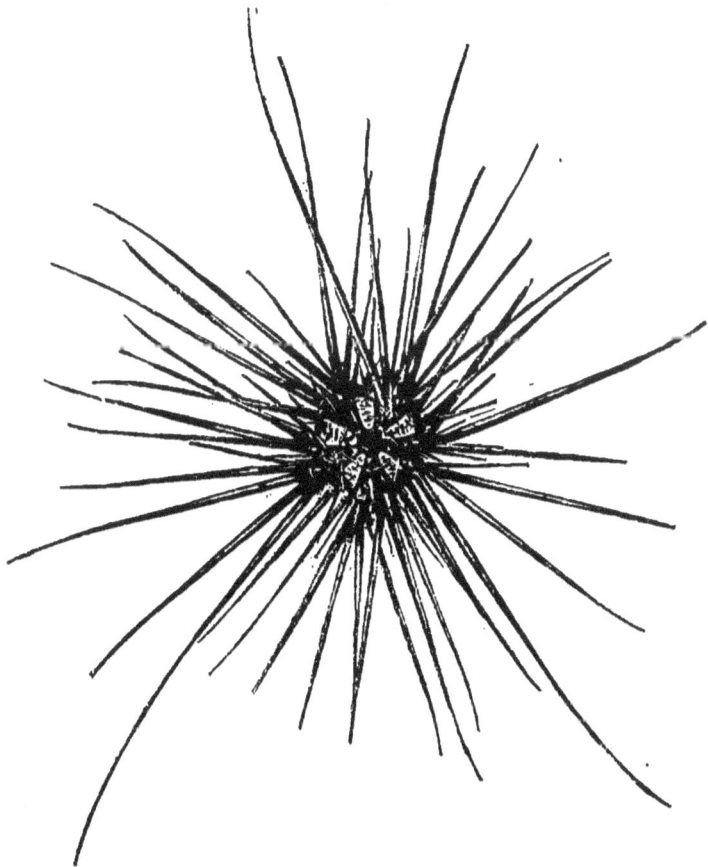

Fig. 358. — Cœlopleurus floridanus. ½.

The primary radioles vary greatly in color, from a delicate straw, often nearly white, to a bright carmine or orange; the base of the spines being usually colored, and the shaft more or less irregularly banded.

The oldest known sea-urchins belong to the Palæchinidæ, a group of palæozoic echini, having, unlike their modern congeners, more than two rows of plates in each zone of the test, and with plates overlapping like the tiles of a roof, so that the test must have possessed considerable flexibility. These urchins were succeeded in mesozoic times by types with a still more flexible test, the coronal plates forming a continuous series from the mouth to the apical system without the usual sharp distinctions of actinal, coronal, and apical systems. This group is represented in our seas by the Echinothuriæ. We may call attention to the characteristic genus Asthenosoma, belonging to the type of echini with flexible test and overlapping plates (Fig. 359 a), first described by Grube from a single specimen, and subsequently collected by the "Challenger." Grube did not, however, recognize the great importance of his discovery, and it was not until Thomson and Pourtalès dredged these flexible urchins that their affinity to the Echinothuriæ of the chalk and to the Palæchinidæ became evident. Traces of this overlapping of the coronal plates can still be detected in the most specialized of the recent sea-urchins.

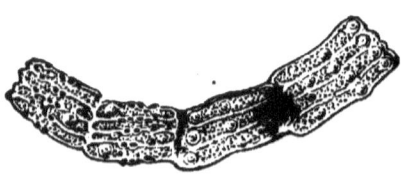

Fig. 359 a. — Asthenosoma hystrix.

In one of the hauls taken between Cape Maysi and Jamaica (1,200 fathoms), we obtained the first specimens of Asthenosoma (Fig. 359) I had seen alive. I was much astonished to find them, fully blown up, hemispherical or globular in shape. This was the shape they always took in subsequent hauls, and on several occasions, when they were obtained from comparatively shallow water near the 100-fathom line, they came up alive, and retained their globular outline. The alcoholic specimens I had seen in the "Challenger" collection dredged from deep water were as flat as pocket-handkerchiefs, and were naturally regarded as flat sea-urchins, although of course endowed with great mobility of test.

Thomson speaks of the vermicular movements passing through the test of Asthenosoma when it assumed on deck what appeared

CHARACTERISTIC DEEP-SEA TYPES. — SEA-URCHINS. 95

to be its normal form and attitude. It is quite dangerous to handle these specimens when alive, the wounds they inflict with

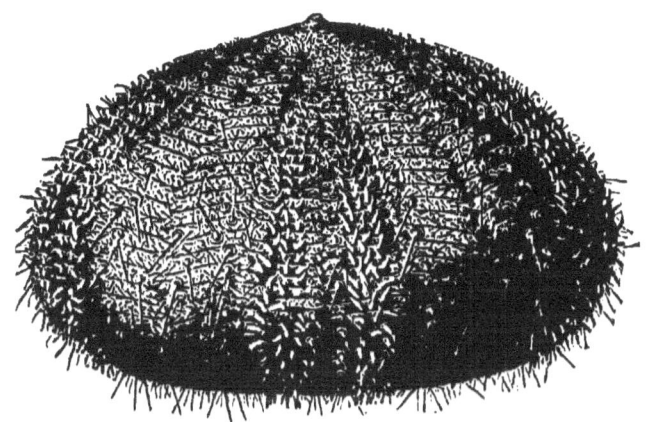

Fig. 350. — Asthenosoma hystrix. ⅓.

their numerous minute sharp stinging spines producing a decidedly unpleasant sensation, accompanied by a slight numbness. The sting is fully as painful as that of Physalia. These modern Echinothuriæ were subsequently found very abundant at moderate depths. The "Challenger" dredged a gigantic species from this group, measuring no less than 312 mm. in diameter.

The spines of the lower surface are shod with a peculiar hoof-shaped tip; on the test are sheathed spines unlike those of any

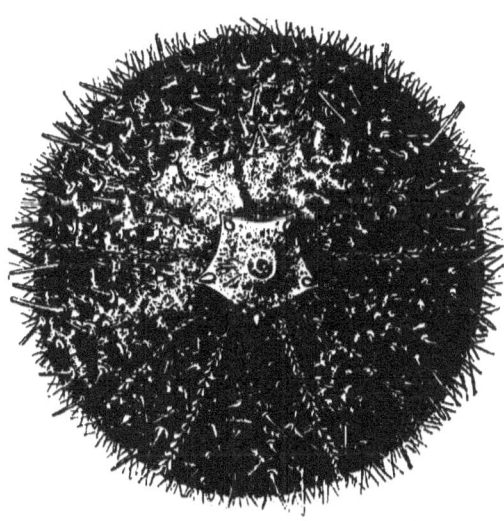

Fig. 360. — Phormosoma placenta. ⅓.

other sea-urchins; they are probably modified pedicellariæ. The test of Asthenosoma is of a deep claret-color. *Phormosoma placenta*, another of the modern Echinothuriæ (Fig. 360), is grayish, or sometimes of a deep brick-color or a yellowish orange. The coronal plates of both zones, although they appear at first glance similar in structure to those of the regular sea-urchins, yet are frequently split up into four distinct plates, as in the palæozoic Archeocidaris and the like.

In a type recalling the Cidaridæ and the Diadematidæ, *Aspidodiadema antillarum* (Fig. 361), remarkable and interesting pedicellariæ are found scattered over the whole of the abactinal part of the test. These may be called sheathed pedicellariæ. The shaft consists of a long, slender radiole, distinctly articulated, surrounded by a huge fleshy sheath, swelling out into three large bags on the sides. (Fig. 362.) The sheath

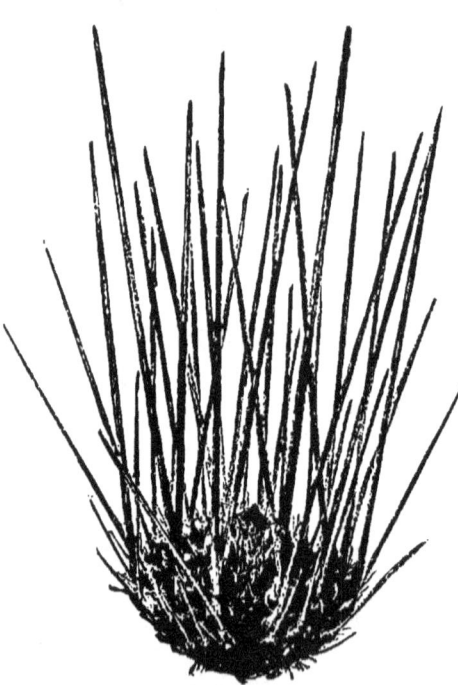

Fig. 361.—Aspidodiadema antillarum. ♀.

Fig. 362.—Aspidodindema antillarum, magnified pedicellaria.

expands at the extremity into a three-lobed cupuliform tip. These pedicellariæ recall the remarkable sheathed secondary spines of Asthenosoma, and form an additional link in the chain of proof that pedicellariæ are merely modified spines. The only other striking genus among the regular urchins is that of Hemipedina (Fig. 363), the modern representative of a family once greatly developed in the cretaceous period.

Although the line to the eastward of Charleston, S. C., was commenced off the very home of the Scutellæ and other clypeastroids, it is remarkable that not a single Mellita or Clypeaster was dredged either on that line or the line run in the axis of the Gulf Stream as far as Cape Hatteras.

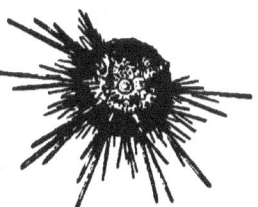

Fig. 363. — Hemipedina cubensis. ¹⁄₁.

We had a similar experience while dredging near the 100-fathom line when approaching the South American continent. The clypeastroids are evidently shallow-water types, with the exception of Echinocyamus, which extends into deep water (805 fathoms), and Echinarachnius, living specimens of which have come up in the trawl from a depth of 524 fathoms off George's Bank. An immense number of dead tests of *Echinocyamus pusillus* were dredged in the Caribbean, the Gulf of Mexico, and the Straits of Florida.[1]

The Nucleolidæ, to which Neolampas (Fig. 364), Rhynchopygus (Fig. 365), and Conolampas belong, are but scantily rep-

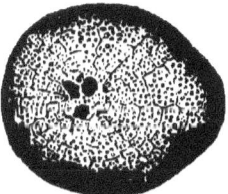

Fig. 364. — Neolampas rostellata. ¹⁄₁.

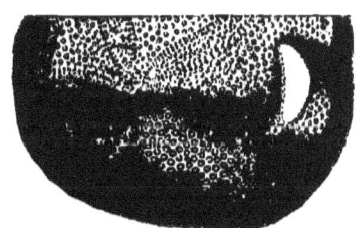

Fig. 365. — Rhynchopygus caribæarum. 2·²⁄₆.

resented in the echinid fauna of to-day. They were once among the most numerous of the urchins, and flourished especially

[1] It is interesting to note, in connection with this, that dead tests of species of Clypeaster, of Echinanthus, of Encope, of Schizaster, of Macropneustes, of Tox- opneustes, of Trigonocidaris, of Temnechinus, of Salenia, and of Cidaris, were also frequently dredged, and sometimes in considerable numbers. This has an

during the cretaceous and jurassic periods. They are, with the Pourtalesiæ, the forerunners of the true spatangoids. They have many features in common with the flat clypeastroids, such as their tuberculation, the character of their pedicellariæ and spines, and the structure of the apical system (Fig. 366), while the structure of the anal system and the general facies of the test rather allies them to the true spatangoids. But neither the Nucleolidæ nor the Pourtalesiæ are possessed of fascioles, an eminently spatangoid structure. These specialized bands of minute spines are slightly developed in some of the cretaceous genera, and their rudimentary form exists to-day in such types as Hemiaster. Their exact function is not yet known. They take their greatest developments in such modern genera as Schizaster. Some light has been thrown on their development by the discovery of a deep-sea species of Macropneustes, which shows a gradual transition between the tuberculation of the test (Fig. 367) and specialized areas corresponding to fascioles.

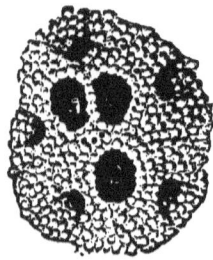

Fig. 366. — Neolampas rostellata, magnified.

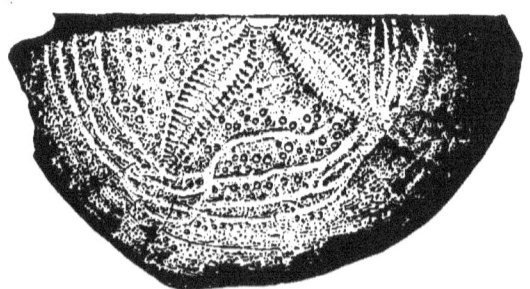

Fig. 367. — Macropneustes spatangoides. ¾.

important bearing as indicating the species which are likely hereafter to be preserved as fossils, and shows us how difficult it may become, even when we have such an abundant and characteristic echinid fauna as that of the West Indies, to reconstruct it from the future fossils. We may also notice that the genera of which we so frequently find the dead tests are the same which have been known as characteristic of the West Indies since the earliest tertiary. We cannot expect to find represented among the fossils the Echinothuriæ, Pourtalesiæ, and many of the Echinidæ, since after death they readily fall to pieces, and may then be dissolved, like many species of mollusks, at great depth, before they become protected by a covering of deep-sea ooze.

The genus Rhynchopygus appeared at the time of the chalk, and is an interesting West Indian type. It is found on both sides of the Isthmus of Panama, and is characteristic of a period when there was a direct connection between the Caribbean Sea and the Bay of Panama.

The allied Neolampas has no fossil representative. The allies of Conolampas date back to the cretaceous period. *Conolampas Sigsbei* (Fig. 368) is by far the most striking sea-urchin I have seen. I shall always remember the particular haul, on the edge of the Yucatan Bank, when the dredge came up containing half

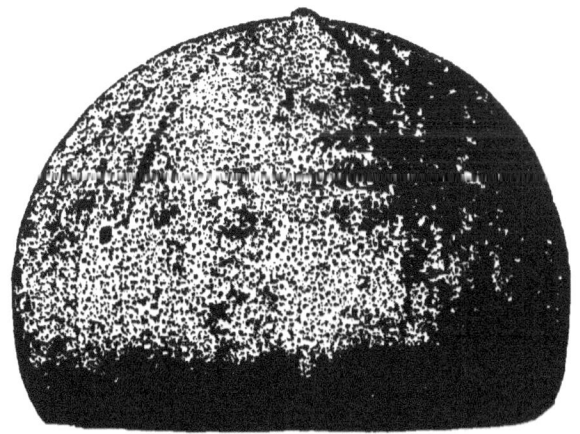

Fig. 368. — Conolampas Sigsbei. ½.

a dozen of these huge brilliant lemon-colored echini. This magnificent species was originally referred to the fossil genus Conoclypus; but Zittel having discovered that some species of this genus possessed teeth, De Loriol made an examination of the genus, and found that it really contained two generic types, one edentate, the other provided with teeth. These discoveries led me to make a renewed examination of *Conoclypus Sigsbei*. On opening a specimen I found that it was edentate. This structural feature is most interesting, as it seems to show us the direct passage, as it were, between the edentate echini and those provided with teeth.

Another typical genus from the chalk represented among the

deep-water spatangoids is Hemiaster (Figs. 369, 370), a small globular genus representing the earlier forms of spatangoids

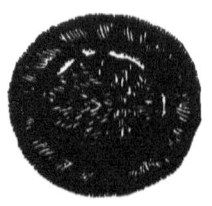 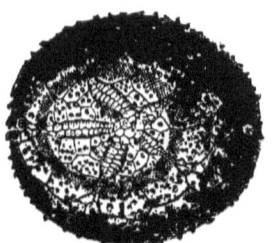

Fig. 369. — Hemiaster zonatus. ¼.   Fig. 370. — Hemiaster expergitus. ?. (Lovén.)

characterized by a simple fasciole. Others belonging to the distinctly cretaceous family of Ananchytidæ are the huge violet or

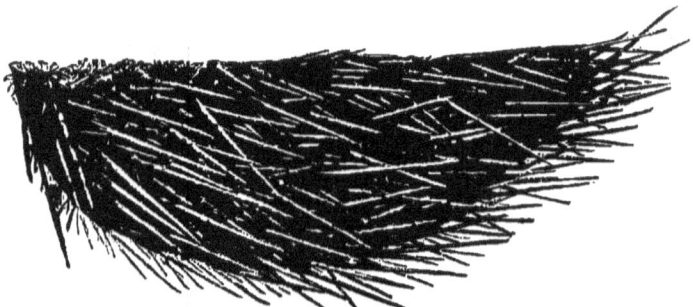

Fig. 371. — Paleopneustes hystrix. ⅔.

deep claret-colored [1] Paleopneustes, characterized by their eminently spatangoid mouth, by their simple ambulacral system and somewhat clypeastroid or even echinidal spines, as in *Paleopneustes hystrix* (Fig. 371), in which the spines resemble more those of a true Echinus than a spatangoid.

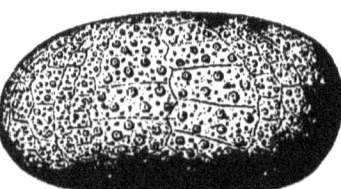

Fig. 372. — Palæotropus Josephinæ. ?.

Other typical modern Ananchytidæ are the West Indian Palæobrissus and Palæotropus. (Fig. 372.)

[1] The colors of the deep-sea echini and other echinoderms seem to be specially fugitive, and greatly discolored the alcohol in which they were placed. The color of the littoral or shallow-water species is far more permanent.

Perhaps the most interesting group of sea-urchins discovered by the late deep-sea explorations are the Pourtalesiæ. The first Pourtalesia was dredged by Pourtalès in the Straits of Florida, — a single specimen only (Fig. 373), but sufficiently perfect to enable me to make an examination of this extraordinary type, so different at first glance from any sea-urchin previously known.

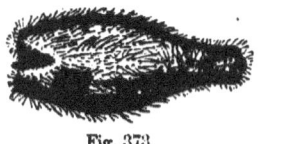
Fig. 373.

Fig. 374.

Pourtalesia miranda. ?.

The study of that species, *Pourtalesia miranda* (Fig. 374), showed affinities to a singular family of urchins described from the chalk, as well as extended relationship to types considered as long extinct. The Ananchytidæ, to which the Pourtalesiæ are allied, are perhaps the most typical cretaceous sea-urchins. They all have large coronal plates, recalling the Echini, with a disconnected apical system characteristic of many cainozoic spatangoids; they have a sunken anal system, some of them a most remarkable anal beak, and a very striking pouch, in which the mouth is placed. They possess rudimentary fascioles, and their

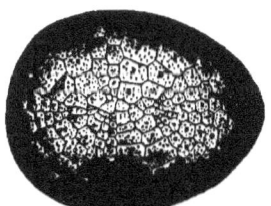

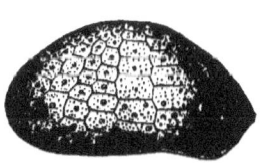

Fig. 375. — Urechinus naresianus. ⅟₁⁵.   Fig. 376.— Profile of Fig. 375.

tuberculation allies them to the clypeastroids. Another species of the same group, which has a wide geographical distribution, is *Urechinus naresianus* (Figs. 375, 376), which seems to be as common in some parts of the Pacific as in the Atlantic.

STARFISHES.[1]

The predominant species of starfishes belong to the Goniasteridæ and Archasteridæ, families which seem thus far to have flourished principally during the chalk. Not only is the number of species belonging to these families very great, but also the number of specimens brought up by the dredge. For instance, *Archaster mirabilis* (Fig. 386) came up by hundreds; it is

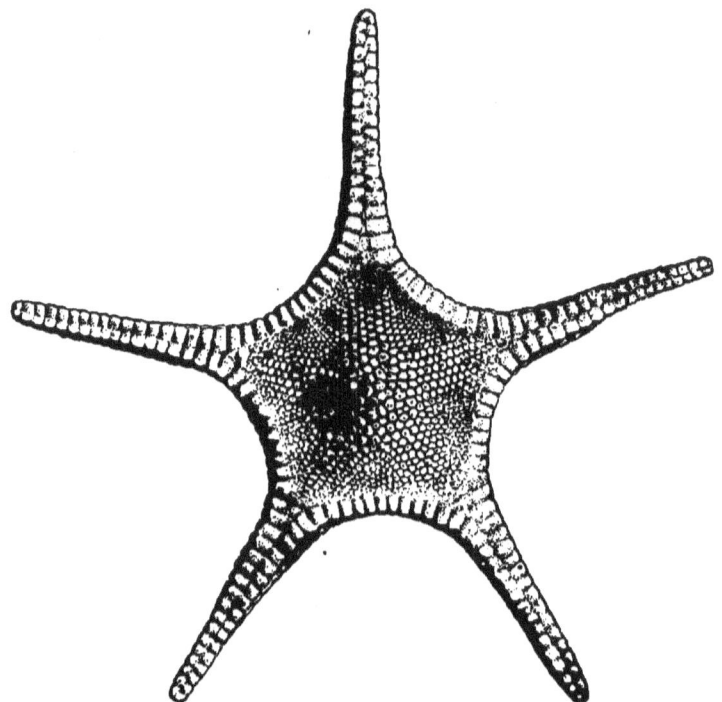

Fig. 377. — Pentagonaster ternalis. ⅔. (*Perrier.*)

a most variable species, extending from 56 fathoms to a depth of 1,920 fathoms. The species of Archaster are common in all depths of the Atlantic, and their number is great. Among the

[1] The account of the Starfishes collected by the "Blake" is compiled from the Report of Professor Perrier, published in the "Nouvelles Archives du Muséum."

novelties described by Perrier, the species of Goniopecten reveal many points of similarity in the structure of Pentagonaster (Fig. 377), Archaster (Fig. 378), and Astropecten, which were all supposed to be radically distinct. The genus Anthenoides (Fig. 379) is intermediate between Anthenea, with large pedicellariæ, and Pentagonaster, with smaller ones and granules. Ctenaster (Fig. 380), on the other hand, recalls a gigantic Ctenodiscus without ventral scales; its marginal plates ally it to the Goniasteridæ, and the structure of its dorsal skeleton to such genera as Solaster and Acanthaster. Radiaster (Fig. 381), a large five-armed starfish, with bunches of spines like *Solaster endeca*, of which it possesses the mar-

Fig. 378. — Archaster pulcher. ¼. (Perrier.)

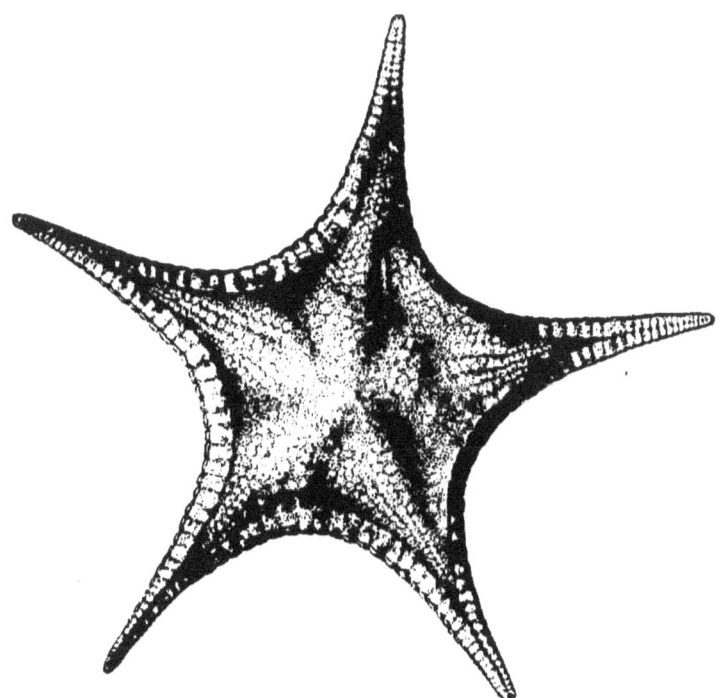

Fig. 379. — Anthenoides Peircei. ⅔. (Perrier.)

ginal plates, its ventral plates allying it to the Asterinæ, finds its place between it and the Astropectinidæ.

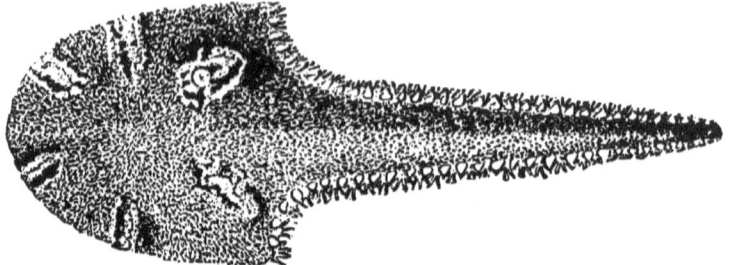

Fig. 380. — Ctenaster spectabilis. ⅔. (Perrier.)

In regard to the geographical distribution of starfishes, it was interesting to find in the deep waters of the Caribbean district the Northern Atlantic genera Cribrella, Solaster, Pedicellaster, and Brisinga.

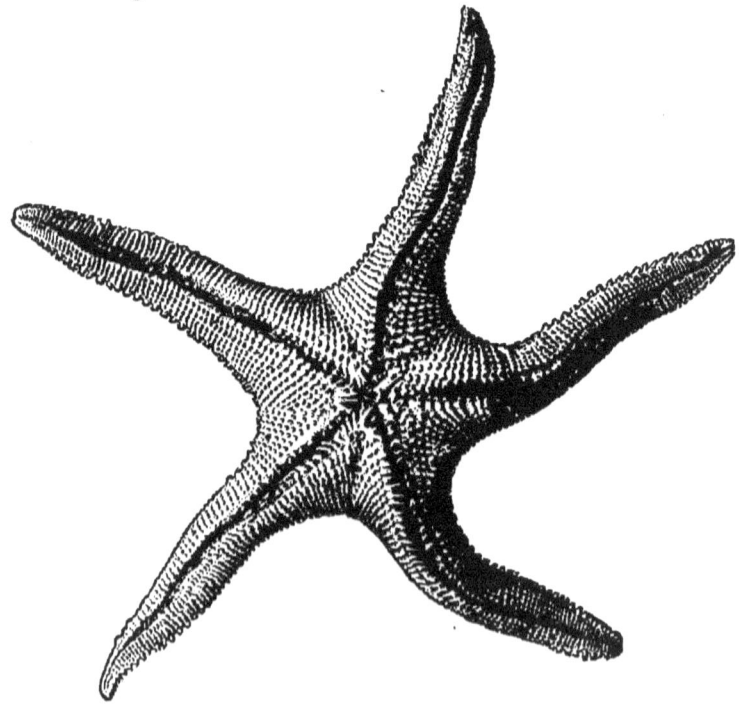

Fig. 381. — Radiaster elegans. ⅔. (Perrier.)

In 1874, Sir Wyville Thomson described Zoroaster, discovered by the "Challenger,"—a genus remarkable for the thickness and regularity of the skeleton. The "Blake" dredged two interesting species of this genus; the one, *Zoroaster Sigsbeei*, with large ossicles of the disk and most distinct arms; the other, on the contrary, *Zoroaster Ackleyi* (Fig. 382), with arms and disk united, giving it an external resemblance to Chætaster, the plates of the actinal surface being crowded with small flattened spines, recalling Luidia, the tentacles in four rows at the base and two rows at the tip ending in a minute disk.

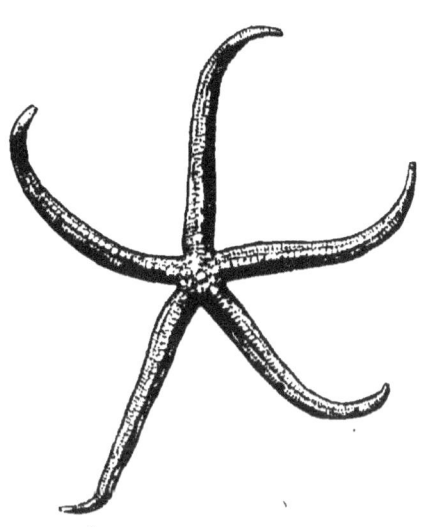

Fig. 382.— Zoroaster Ackleyi. ¾. (Perrier.)

Hymenodiscus Agassizii (Fig. 383) belongs to an intermediate type far more pronounced even than Brisinga. It recalls the ophiurans by its round disk, distinctly separated from the arms, which are long, slender, and mobile, furnished with a lateral row of spines, as in the ophiurans, which may serve as organs of locomotion. But there are twelve arms in these starfishes, while there are not more than six, or sometimes eight, in ophiurans. The disk is membranous (Fig. 384), with a circle of ossicles formed from the first joint of the arms. The skeleton of the arms is most simple, consisting of four longitudinal series of pieces; each piece carries a long lateral spine (Fig. 385), covered by a smooth sheath swollen at the extremity, and a cluster of pedicellariæ such as characterize the starfishes. The true starfish ambulacral pieces are wanting in Hymenodiscus. The dorsal skeleton of Brisinga may be considered as only a shield of the genital glands, which are similar in their structure, as is the digestive cavity, to the same organs of the ophiurans, while the structure of the ambulacral furrow approaches that of the Comatulæ. Hymenodiscus and Brisinga thus form

Fig. 383.—Hymenodiscus Agassizii. $\frac{1}{1}$ⁱ. (Perrier.)

among starfishes a very peculiar family, marked by most exceptional characters. The study of Hymenodiscus, closely allied

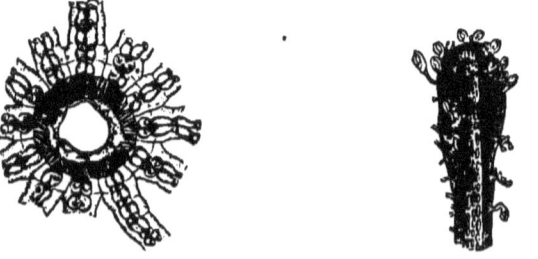

Fig. 384. $\frac{8}{1}$.   Fig. 385. Magnified spine.
Hymenodiscus Agassizii. (Perrier.)

to the northern Brisinga, has had an important bearing on the morphology of the starfish skeleton.

There exists in *Archaster mirabilis* (Fig. 386) a remarkable sort of pedicellariæ, consisting of two ossicles placed face to face like the hooks of a bracket, each carrying a comb of spines falling one towards the other and forming a very complicated organ of prehension.

There seems to be no doubt that the starfish fauna becomes less and less varied as the depth increases, the maximum development in individuals being found at a depth of from 100 to 250 fathoms. The number of species does not seem to diminish so rapidly as the number of individuals, nor in proportion to the variation of the nature of the bottom.

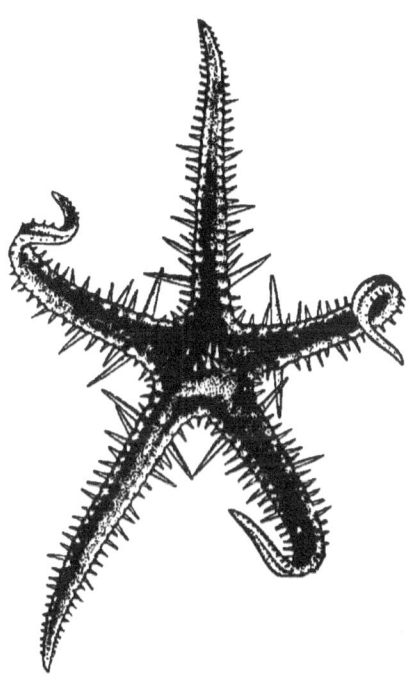

Fig. 386. — *Archaster mirabilis*. ¼. (Perrier.)

Thus in depths of less than 100 fathoms it required 2.7 hauls of the dredge to bring up one species, 15 species and 150 specimens being collected in 41 hauls. Between 100 and 200 fathoms, 21 species and 144 specimens being obtained, the coefficient was 3.6. From 200 to 300 fathoms the coefficient was 3.15, with 13 species and 66 individuals. From 300 to 400 fathoms only 12 individuals were dredged, belonging to 9 species, the coefficient being 3.9. Between 400 and 500 fathoms the coefficient was 4.6. Between 500 and 600 fathoms the coefficient had become 13. We made 15 hauls between 800 and 900 fathoms, but obtained only 3 species and 3 individuals, although at a depth of 1,900 to 2,000 fathoms, 4 hauls gave us 7 specimens of 4 species.

Of course the method of carrying on dredgings affects the

results to a certain extent. The greater length of time required for dredging in considerable depth and the state of the weather

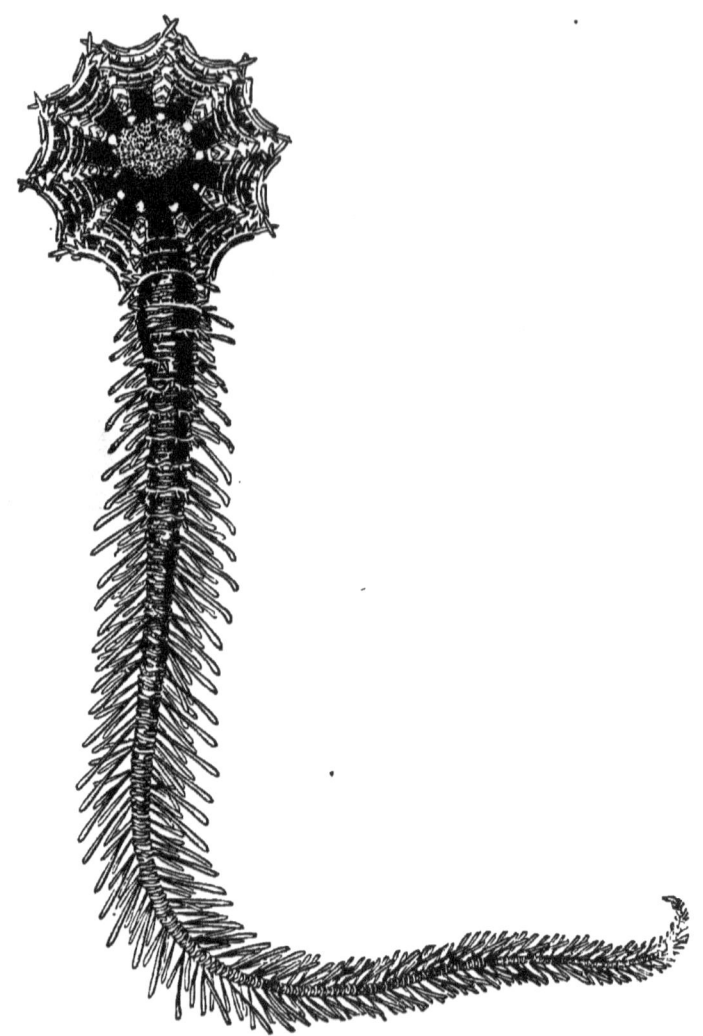

Fig. 387. — Brisinga coronata. ⅔. (Sars.)

while at specially favorable localities naturally influence the success in collecting.

CHARACTERISTIC DEEP-SEA TYPES. — OPHIURANS.

North of Cape Hatteras the species of starfish procured by the "Blake" are identical with those described by Professor Verrill from the dredgings of the United States Fish Commission.

I would only mention here, as the most interesting of the species we found, ten-armed specimens of *Brisinga coronata*. (Fig. 387.) This species forms the subject of an elaborate paper by the younger Sars, and I have here reproduced one of his figures.

OPHIURANS.[1]

Among Echinoderms there are two families, the brittle-stars, or Ophiuridæ, and the branching-stars, or Astrophytidæ, which are distinguished by a peculiar axis in the arms, made up of articulated bones somewhat like vertebræ. The disk or body is

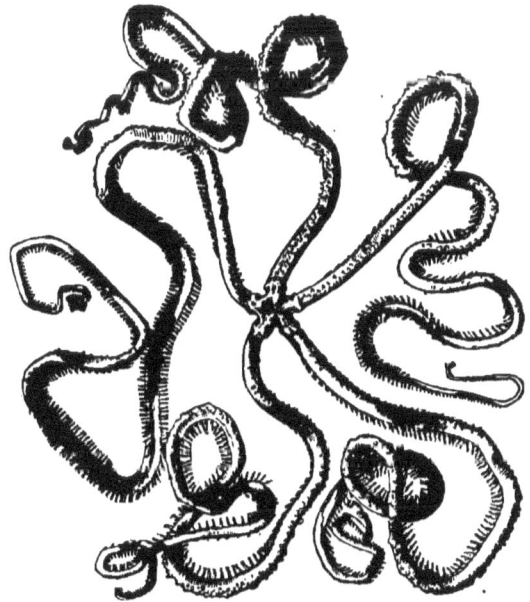

Fig. 389. — Ophiocreas spinulosus. ½.

usually distinctly set off from the arms. These last contain no prolongation of the central digestive cavity, as they do in the starfishes proper.

[1] Mr. Lyman has prepared the account of the ophiurans.

Besides the peculiarity of branching arms (*Astrophyton cœcilia*, Fig. 388) which distinguishes some of the genera, the Astrophytidæ have characteristic joints in the arm-axis, which separate them from the Ophiuridæ. They are also usually covered, not by conspicuous plates of lime carbonate, but by a leathery skin (*Ophiocreas spinulosus*, Fig. 389). The typical Ophiuridæ have a well-marked central disk covered with plates or scales, and from it radiate five arms encased in four longitudinal rows of plates (*Ophiozona nivea*, Fig. 390). The side arm-

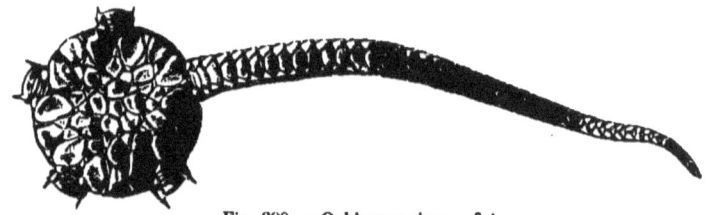

Fig. 300. — Ophiozona nivea. ⅔.'

plates bear spines, which may lie close along the arm (*Ophiophyllum petilum*, Fig. 391), or stand out from it at a strong

Fig. 301. — Ophiophyllum petilum. ½.

angle (*Ophiocamax hystrix*, Fig. 392). There is an almost endless variety in the shape, consistency, number, and size of

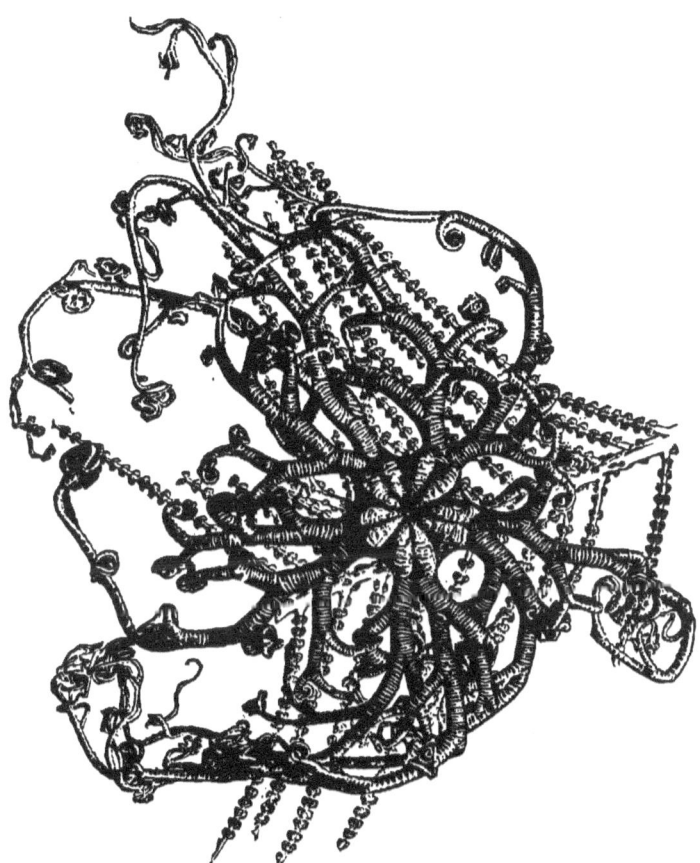

Fig. 388. — Astrophyton cœcilia. ½.

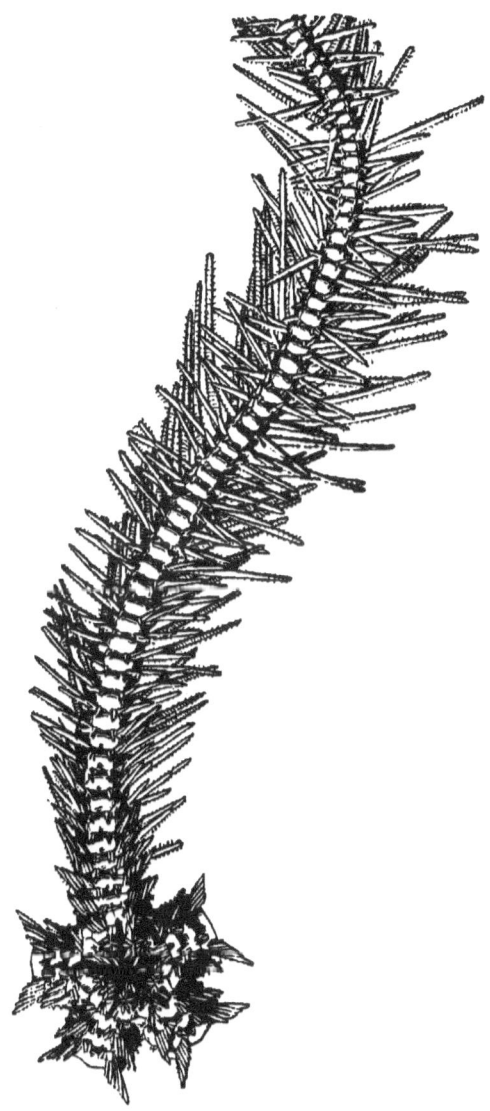

Fig. 302. — Ophiocamax hystrix. ⅔.

CHARACTERISTIC DEEP-SEA TYPES. — OPHIURANS. 111

the plates, scales, spines, and granules (*Ophiopæpale Goësiana*, Fig. 393; *Ophiura Elaps*, Fig. 394; *Ophioconis miliaria*,

Fig. 393. — Ophiopæpale Goësiana. ⅓.

Fig. 395). So there may be every diversity of general form, from the smooth simple *Ophiomusium planum* (Fig. 396) to

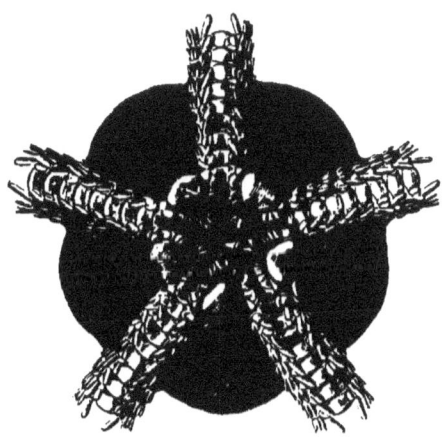

Fig. 394. — Ophium Elaps. ⅘.

the highly complex *Ophiomyces frutectosus*. (Fig. 397.) No conditions could well be more favorable to ophiurans than those of the West Indian waters. A tropical sun gives to the shallows a temperature of over 80° Fahr., which decreases gradually with the depth, till, between 600 and 700 fathoms, it has fallen to

Fig. 395. — Ophioconis miliaria. $\frac{6}{1}$.

391°. The American continents on one side, and the Antilles

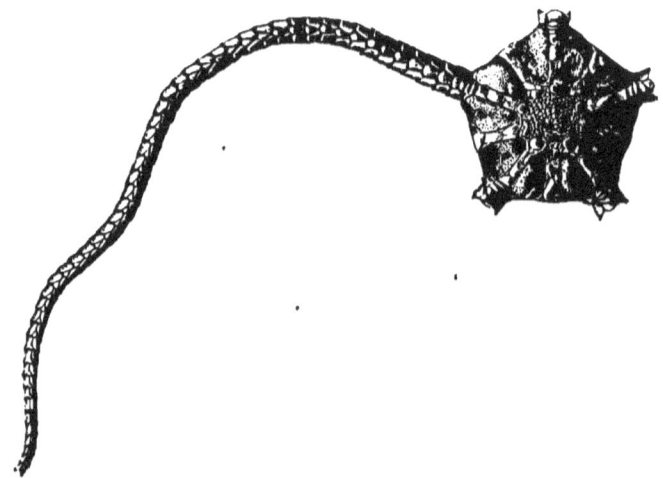

Fig. 396. — Ophiomusium planum. $1\frac{1}{1}$.

on the other, furnish those great land masses the neighborhood of which seems essential to rich and varied marine life. Their wash increases the already abundant supply of lime, a substance that forms nearly the entire weight of some species (*Ophiomas-*

Fig. 307. — Ophiomyces frutectosus. ⅜.

tus secundus, Fig. 398). These conditions naturally give rise to much variety in form, and to a great abundance of individu-

Fig. 308. — Ophiomastus secundus. ⅝.

als. The nine species mentioned by Müller and Troschel, in 1842, as belonging to this area, have increased to one hundred and fifty-five, which are distributed at various depths. On the flats and reefs, near islands and keys, may be found colonies of Ophiothrix, blue, green, or red, with their translucent thorny arm-spines, and the humble Ophiactis swarming on great sponges; while here and there a yellow or vermilion star marks the soft *Ophiomyxa flaccida*. To the brown gorgonians clings the large Ophiocoma, similar in color; and sometimes a Medusa-head, whose branching arms excited the wonder of old Rondelet, twines about the thicker stems. These and their companions, living in a strong light, and in warm shallow water, present brilliant and well-marked colors. Nor are those that inhabit the dark and cold depths of the ocean always pale; on the contrary, many are of a bright orange or red. They are peculiar, however, in that their colors generally fade in alcohol; and in an alcoholic collection the shallow species may readily be distinguished by their brighter coloration.

Like other marine animals, ophiurans are distributed according to the depth and temperature of the water. About one half of the known species are confined to the zone between low-water mark and 30 fathoms. These include the Medusa-heads (Astrophyton). But not less than one tenth of the known living forms are found entirely below 1,000 fathoms, and of these several, such as *Ophiomusium Lymani*,[1] *Ophiocreas spinulosus*, and *Ophiocamax hystrix*, live in great colonies, just as some of the

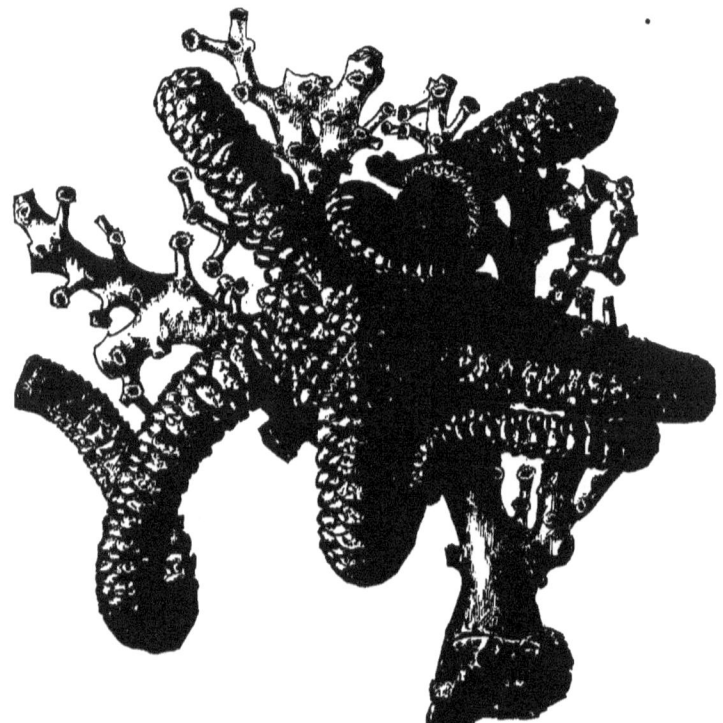

Fig. 399. — Sigsbeia murrhina. ½.

shallow-water species do. They are found in various situations. The localization of some of these is very marked. The stiff-armed *Ophiomusium* of deep water, with their swollen tuberculous plates, naturally lie on the bottom, while other species with supple or prehensile arms, such as Sigsbeia (Fig. 399) and As-

[1] All the way from Cape Hatteras to the extremity of George's Bank, *Ophiomusium Lymani* was quite common in deep water.

CHARACTERISTIC DEEP-SEA TYPES. — OPHIURANS. 115

trocnida (Fig. 400), cling to corals, gorgonians, and bryozoans. One species, *Ophiomitra valida*, is often found twined round the stalk of a sea-lily (Pentacrinus). Off the mouth of the

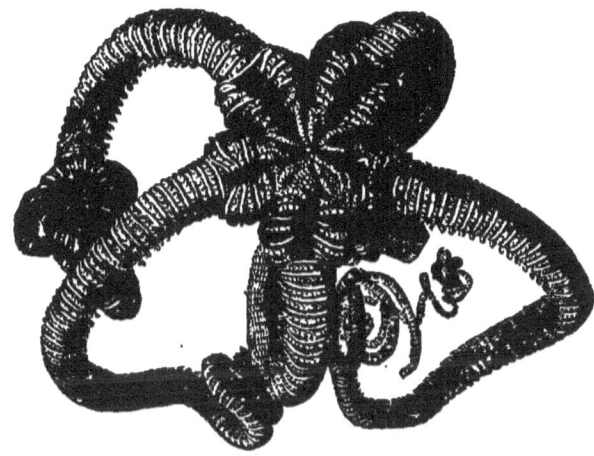

Fig. 400. — Astrocnida isidis. ½.

Mississippi we brought up from 100 fathoms a number of *Ophiolipus Agassizii* (Fig. 401), which must live buried in the mud

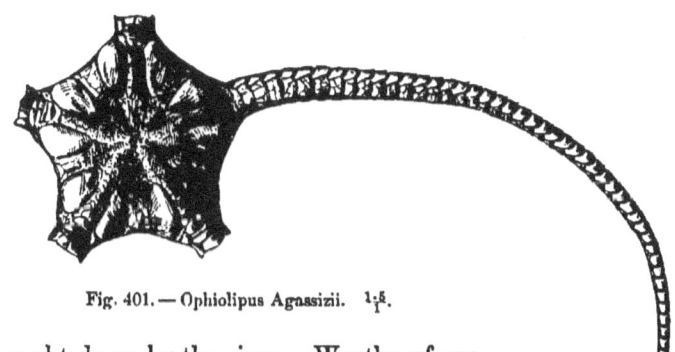

Fig. 401. — Ophiolipus Agassizii. 1⅝.

brought down by the river. Worthy of special notice is a small soft ophiuran which came up in the very last cast made by the "Blake" off Barbados. This seemed to have little tufts resembling bunches of hydroids on the sides of the arms (Fig. 402):

further examination showed these tufts to be bunches of minute spines enclosed each in a thick skin bag, resembling long-stemmed parasols with small shades. This structure differs radically from that of the spines of all other ophiurans hitherto known where there is no departure from the single row of articulated spines.

Fig. 402. $\frac{6}{1}$.
Ophiohelus umbella.

Fig. 403. $\frac{40}{1}$.
(Lyman.)

A bunch of these umbrella-shaped spines of *Ophiohelus umbella* is given in Fig. 403.

CRINOIDS.[1]

The stalked crinoids are among the most interesting of the deep-sea animals. Their palæontological relations run back in the case of the Pentacrinoidea and the Apiocrinidæ (Rhizocrinus) to the jurassic period; while the relationship of Holopus may probably extend to the silurian (Edriocrinus).

The Pentacrinidæ, of which four species were known from the Caribbean district, are characterized by the verticillate arrangement of the cirri along the whole length of the stem, while in the Bourgueticrini the whole stem even may be free of cirri. Recent species of Pentacrinidæ have been found both in the Pacific and Atlantic, and they are common at depths of less than 100 fathoms. The species of the genus Metacrinus (Fig. 404) replace in the Pacific, to a certain extent, the Atlantic Pentacrini. Our first accurate knowledge of the type dates from Miller, who compared the structure of the fossil species with that of both *Pentacrinus asterius* (Fig. 405) and the free Comatulæ. This relationship was subsequently most satisfactorily proved by J. V. Thomson, who in 1836 discovered the pentacrinoid stage of a species of Comatula. (Fig. 406.)

There seems to be no special order in the division of the secondary and tertiary arms of the Pentacrinidæ, though the dif-

---

[1] The account here given of the Crinoids is drawn up from the Reports made by Dr. P. H. Carpenter on the collections of the "Challenger" and "Blake" expeditions.

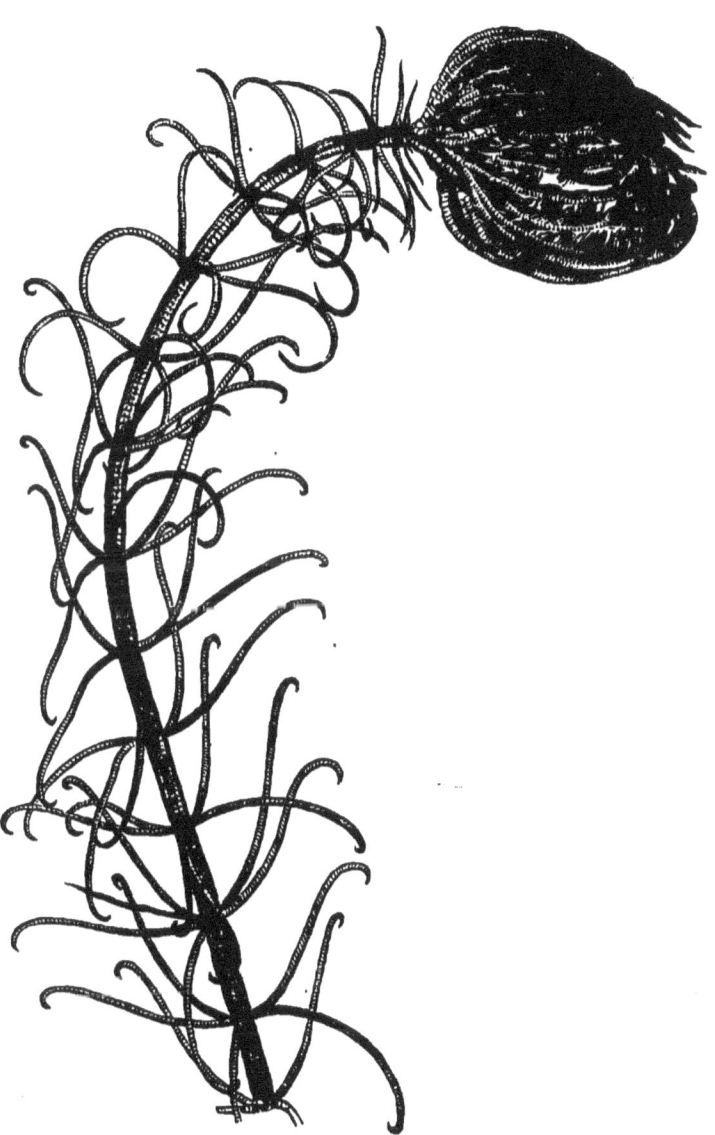

Fig. 405. — Pentacrinus asterius. ½. (Carpenter.)

CHARACTERISTIC DEEP-SEA TYPES. — CRINOIDS. 117

Fig. 404. — Metacrinus angulatus. ½. (Carpenter.)

Fig. 406. — Pentacrinus stage of Actinometra meridionalis, magnified.

ferences existing between the liassic and the modern Pentacrinidæ led the Austins to establish the genus Extracrinus for a fossil species which seems to have been gregarious, and of which Queenstedt has figured a magnificent slab, some of the specimens with a stem nearly sixty feet long. The stems were often twisted into a solid, ropelike mass, and are so entangled on the slab that it is difficult to make out the individual stems. A similar entangling also occurred among the specimens dredged by the "Blake," and it was often very difficult to separate specimens the cirri of which had become attached to adjoining stems. It is possible that they may live gregariously, more or less united either by the twisting of the stems or the grappling of the cirri, and be only loosely attached to the ooze on which they live, or anchored more securely by the terminal whorl to some projecting piece of rock or gorgonia stem.

Crinoids both stalked and free live in colonies. Comatulæ are most abundant in certain localities. *Antedon Sarsii* was brought up in thousands by the "Blake." The U. S. Fish Commission and the "Challenger" have had a similar experience with different species of Comatulæ. On one occasion, off Sand Key, we must have passed over a field of Rhizocrinus with the dredge, judging from the number of stems and heads of all sizes it contained. The oldest species known, *Pentacrinus asterius* (Fig. 405), is marked by its greatly multiplied large and strong arms, while in *P. decorus* (Fig. 407) the number is greatly reduced. We know but little of the young of Pentacrinus. The youngest specimens dredged by the "Blake," and figured by Carpenter (Fig. 408), show the great relative height of the stem joint as a characteristic feature of young specimens. The stems of *Pentacrinus asterius* and *P. decorus* are longer than those of the other species of the genus. *P. Mülleri* (Fig. 409) was discovered by Oersted, and in 1865 Dr. Lütken gave a detailed account of the West Indian Pentacrinidæ; the many specimens of Pentacrinus dredged by the "Blake" were originally identified with it, but, as has been clearly shown by Carpenter, they all belong to *P. decorus*. Both *P. Mülleri* and *P.*

Fig. 408. — Pentacrinus decorus. ⅔. (Carpenter.)

Fig. 407. — Pentacrinus decorus. ⅔. (Carpenter.)

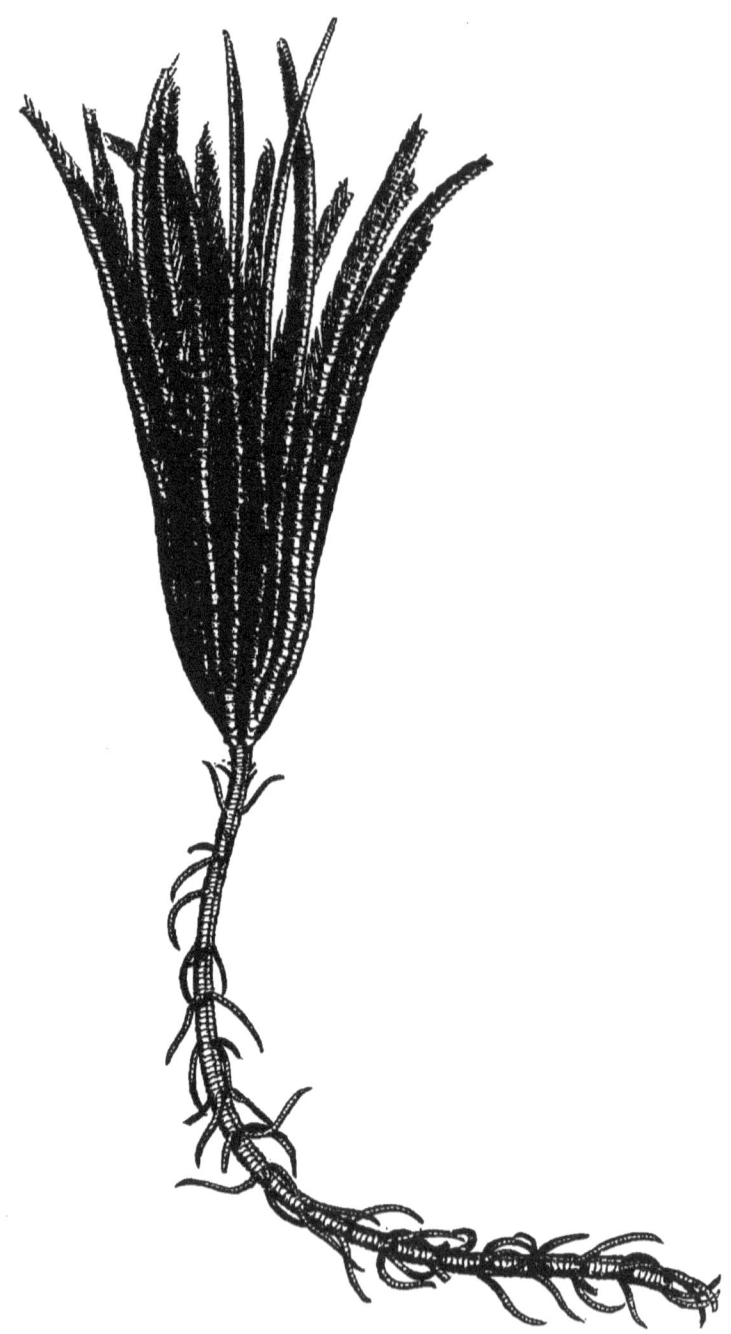

Fig. 410. — Pentacrinus Blakei. ¾. (Carpenter.)

decorus seem most variable species of a very variable genus. Off St. Vincent the specimens brought up evidently lived on a rocky bottom, and there the specimens were undoubtedly anchored by the terminal cirri, their stems having become fractured, as has been suggested by Thomson, at the nodes. Thus they continued to lead a semi-free existence, the lowest nodal joint becoming smooth and rounded, showing that the animal had been free for some time, the nodal terminal joint being surrounded by its whorl of cirri, which curved downward like a grappling-iron,[1] so that the animal must have been able to change its position at pleasure by swimming with its arms, like Comatulæ. Another species of Pentacrinus obtained has been named *P. Blakei* by Dr. Carpenter. (Fig. 410.) It has a slender,

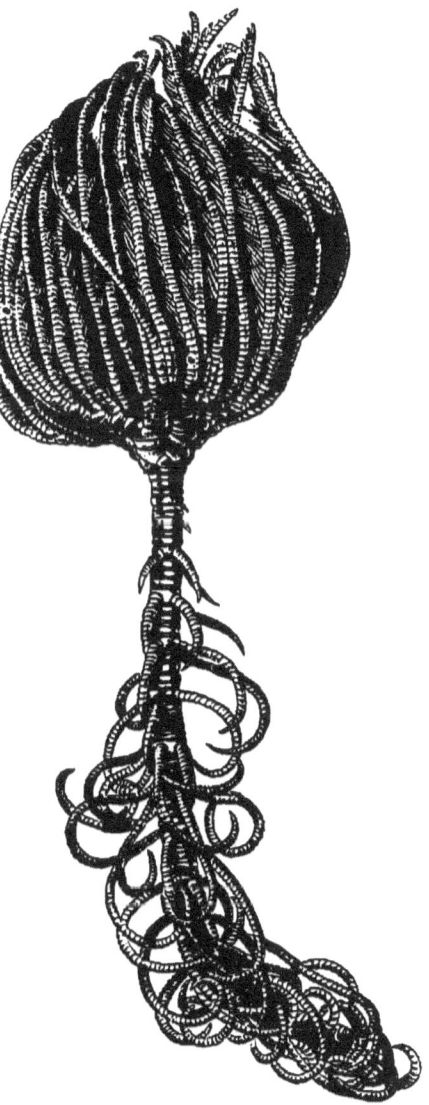

Fig. 409. — Pentacrinus Mülleri. ⅔. (Carpenter.)

[1] In regard to the movements of Pentacrinus the following extract from a letter of Lieut.-Commander C. D. Sigsbee will be of interest : —

"On the 1st of April we put to sea again [from Havana] ; we steamed about one and a half miles from the Morro (east), and at the third haul, in 177 fath-

smooth stem, with a rounded pentagonal outline; it is apparently not common, having been dredged by the "Blake" only at four localities.

Rhizocrinus (Fig. 411) has a stem composed of dice-box shaped joints, terminating in a spreading root or a number of branching radicular cirri, not arranged in definite whorls, with a high calyx. It was first named by M. Sars, who afterwards described it, in 1868, as belonging to the Apiocrinidæ. But before the appearance of Sars's memoir, this interesting crinoid had been rediscovered by Pourtalès, and stated by him to belong undoubtedly to the genus Bourgueticrinus of D'Orbigny, and he gave it the provisional name of *B. Hotessieri*, thinking it might prove identical with a crinoid of that name of which fragments had been found in the recent limestones of Guadeloupe. Pourtalès was the first to make out accurately the composition of the cup, and he of course also recognized its identity with the Rhizocrinus of Sars's memoir, *R. lofotensis*. Rhizocrinus has been dredged by the Porcupine, the Hassler, the

oms, from disintegrated coral rock bottom, up came six beautiful 'sea lilies.' Some of them came up on the tangles, some on the dredge. They were as brittle as glass. The heads soon curled over, and showed a decided disposition to drop off. At a haul made soon after we got more, and, being afraid to put so many of them in the tank together, I tried to delude the animals into the idea that they were in their native temperatures by putting them into ice-water. This worked well, although some of them became exasperated and shed some of their arms. They lived in the ice-water for two hours, until I transferred them to the tank. They moved their arms one at a time. Some of the lilies were white, some purple, some yellow; the last was the color of the smaller and more delicate ones."

I have nothing to add to the general description of their movements given by Sigsbee, with the exception of their use of the cirri placed along the stem. These they move more rapidly than the arms, and use them as hooks to catch hold of neighboring objects; and, on account of their sharp extremities, the cirri are well adapted to retain their hold. The stem itself passes slowly from a rigid vertical attitude to a curved or even drooping position. We did not bring up a single specimen that showed the mode of attachment of the stem. Several naturalists, on the evidence of large slabs containing fossil Pentacrini, where no basal attachment could be seen, have come to the conclusion that Pentacrini might be free, attaching themselves temporarily by the cirri of the stem, much as Comatulæ do. I am informed, however, by Captain E. Cole, of the telegraph steamer "Investigator," that he has frequently brought up the West Indian telegraph cable with Pentacrini attached, and that they are fixed, the basal extremity of the stem spreading slightly, somewhat after the manner of Holopus, so that it requires considerable strength to detach them.

CHARACTERISTIC DEEP-SEA TYPES. — CRINOIDS. 121

Challenger, the Blake, the Talisman, and by the U. S. Fish Commission; it has a very wide geographical distribution, hav-

Fig. 412. — Rhizocrinus Rawsoni. ¼. (Carpenter.)

ing been found as far north as the Lofoten, and as far south as 35° south latitude. It is very common in the Gulf of Mexico, the Caribbean, and along the east coast of the United States. It is of a brownish-chestnut color when alive, varying from that to a dirty white. *R. Rawsoni,* a second species (Fig. 412), was first dredged by the "Porcupine" off Cape Clear, but the specimens were considered a variety of the other species,

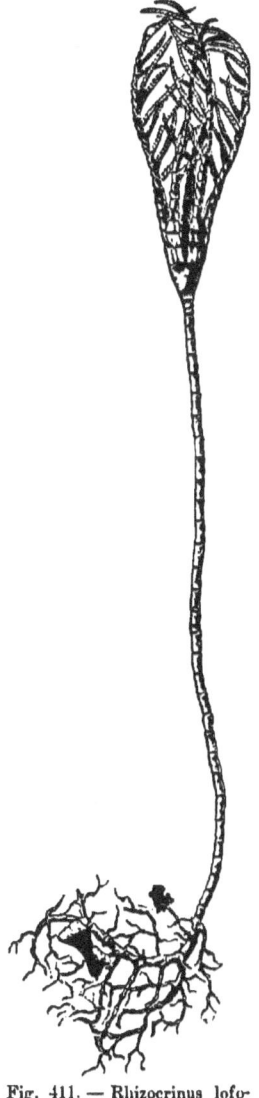

Fig. 411. — Rhizocrinus lofotensis. ²⁄₁. (Sars & Carpenter.)

until attention was again called to it by Pourtalès in 1871, and he showed the stout-stemmed specimens collected by the "Hassler" off Barbados to be of a species distinct from the one previously described.

The predecessors of Rhizocrinus were well represented in the lower tertiary, and go back to the cretaceous, where its ally, Bourgueticrinus, was very abundant.

In Rhizocrinus, spreading rootlets extend below the regular joints. By expansion at the ends or sides of these rootlets the animals attach themselves to any foreign body they happen to find in the deep ooze in which they become anchored; when once fixed, they probably remain so for life. The stem joints of the Bourgueticrinidæ are movable upon one another; they are not uniformly discoidal, like those of the Pentacrinidæ, but are strung as it were upon five tendons of variable length.

Agassiz, who watched the movements of *Rhizocrinus Rawsoni*, says:—

"When contracted, the pinnules are pressed against the arms, and the arms themselves shut against one another, so that the whole looks like a brush made up of a few long coarse twines. When the animal opens, the arms at first separate without bending, but gradually the tip of the arms bends outwards as the arms diverge more and more, and when fully expanded the crown has the appearance of a lily. I have not been able to detect any motion in the stem traceable to contraction, though there is no stiffness in its bearing. When disturbed, the pinnules of the arms first contract, the arms straighten themselves out, and the whole gradually and slowly closes up. It was a very impressive sight for me to watch the movements of this creature, for it told not of its own way only, but at the same time afforded a glimpse into the countless ages of the past, when these crinoids, so rarely seen nowadays, formed a prominent feature of the animal kingdom. I could see, without great effort of the imagination, the shoal of Lockport, teeming with the many genera of crinoids which the geologists of New York have rescued from that prolific silurian deposit, or recall the formation of my native country, in the hillsides of which, also among fossils indicating shoal-water beds, other crinoids abound, resembling still more closely those we find in these waters."

The English, French, and Norwegian expeditions discovered also other stalked crinoids belonging to the genera Bathycrinus,

Hyocrinus, and Ilycrinus; but the " Blake " was not fortunate enough to obtain any of these.

The last and perhaps most interesting of the West Indian stalked crinoids belongs to the genus Holopus. (Figs. 413, 414.) Less than half a dozen specimens of it are known to exist. The first specimen collected is now in the museum of the École des

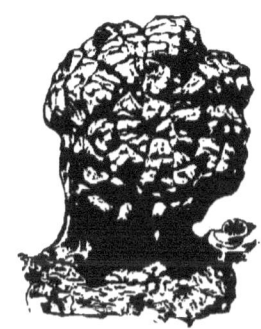

Fig. 413. — Adult Holopus Rangi. ⅓⁹⁄₁. (Carpenter.)

Fig. 414. — Half-grown Holopus Rangi. ⅔. (Carpenter.)

Mines. Sir Rawson W. Rawson, when Governor of Barbados, obtained three specimens, which were lent to Sir Wyville Thomson, and have, with the material of the "Blake," formed the basis of Dr. Carpenter's work on the subject.

The genus Holopus was established by D'Orbigny, in 1837, from a single specimen which was brought from Martinique by Sander Rang. Its true nature was not recognized by other palæontologists, some of whom considered it to be a barnacle. The dried specimens all have a blackish green tinge ; the single arm dredged off Montserrat had a whitish tint. The arms of all the specimens are strongly curved, closing the disk entirely ; but of course this is not the natural attitude of the animal. Holopus is attached by an irregularly expanded base, formed by the extension of the tubular calyx, which is slightly bent, while a constriction separates the cup from the spreading base. The youngest specimen (Fig. 415), of jet-black color, dredged off Bahia Honda, only 3 mm. in diameter, differs very much from the older specimens, as will be seen from the figures. The specimen was attached to a piece of rock, and was not detected until it had become dry. The general shape is a contracted truncated cone, with irregular contour of attachment. The surface is granulated or shagreen-like, with a few small tubercles scattered over it.

The great peculiarity of the Caribbean fauna is the abundance

Fig. 415. — Holopus Rangi. $\frac{1}{1}$.

of ten-armed Comatulæ representing both the principal genera. About two thirds of the Antedon species and three fourths of the Actinometræ belong to this simple type. In this respect the contrast with the Comatula fauna of the Eastern seas is very marked. Ten-armed forms of both genera are there decidedly in the minority.

Of all the Antedon species dredged by the Coast Survey expeditions, that with the widest range within the Caribbean Sea is the little ten-armed *Antedon Hagenii* Pourt. It was obtained by the "Blake" on the Yucatan Bank, and also at various stations between Dominica and Grenada, at different depths between 75 and 291 fathoms; while Pourtalès dredged it in great abundance at several localities in the Straits of Florida. Among the large number of individuals of *Antedon Hagenii* from the Straits of Florida, Carpenter noticed a few examples of two new Antedon species. One of them is distinguished by having enormous lancet-like processes on the lower joints of its oral pinnules; while the other is a very exceptional type, with no pinnules at all upon the second and third brachials, though those of the other arm-joints are developed as usual. This is a singular condition, which occurs but rarely among the Comatulæ. Except in the remarkable type Atelecrinus (Fig. 416), which has no pinnules at all upon the ten or twelve lower arm-joints, these are the only Comatulæ which Carpenter has ever met with, in

Fig. 416. — Atelecrinus. $\frac{2}{1}$. (Carpenter.)

CHARACTERISTIC DEEP-SEA TYPES. — CRINOIDS. 125

an examination of several hundred individuals, that present any departure from the ordinary pinnule arrangement.

The two Comatulæ which from their abundance seem especially characteristic of the neighborhood of the Caribbean Islands, ranging from Santa Cruz to Grenada, are an Antedon and an Actinometra, both of which had been obtained previously to the "Blake" expedition. In the year 1870, Duchassaing brought from Guadeloupe to the Paris Museum a fine specimen of Antedon, with thirty very spiny arms. Carpenter readily recognized it in the "Blake" collection, and has named it *Antedon spinifera*. (Fig. 417.) The common Actinometra of the Caribbean Sea is a singularly protean species, which was obtained at thirty stations by the "Blake." The "Hassler" dredged it off Barbados, and it was found by the "Investigator" off St. Lucia, and also attached to the Martinique and Dominica cable.

Fig. 417. — Antedon spinifera. ¼.

It ranges from 73 to 278, and possibly to 380 fathoms. Not only is it everywhere very abundant, but it presents a most remarkable series of minor variations on one fairly distinct type, which, under the name of *Actinometra pulchella* (Fig. 418), includes no less than six forms apparently distinct at first sight. Most of the specimens have twenty arms, occasionally a smaller number; some, however, have as few as twelve to fifteen. *Actinometra pulchella* is also interesting as furnishing an instance of variation from the ordinary type of five rays. One specimen, like that dredged by the "Challenger," has six rays. It is curious that this variation, which is common in Rhizocrinus, should be so rare among the Comatulæ.

The results of Carpenter's examination of the "Challenger" and "Blake" collections, and of the numerous Comatulæ to which he had access in the various European museums, entirely confirm and extend the conclusions to which he had previously been led respecting the separation of Antedon and Actinometra as distinct generic types. A glance at the skeleton is sufficient to enable one to distinguish the genus.

With another species of Comatula, in 450 fathoms, Pourtalès dredged off Cojima two mutilated specimens belonging to a type

Fig. 418. — Actinometra pulchella. ¼.

of singular interest. This new Comatula may be considered as a permanent larval form; and it is not a little singular to find larval characters persisting in recent Comatulæ. For this remarkable combination Carpenter has established a new genus, which he proposes to call Atelecrinus. (See Fig. 416.)

In conclusion, I may mention that many of the Comatulæ examined were the hosts of Myzostomidæ. These curious parasites have been fully described by Dr. von Graff, from the material of the "Challenger" and "Blake." I give here figures of two characteristic species of these worms, strangely modified to adapt themselves to the peculiar conditions of their habitat.

The organs of the body are arranged radially, and the muscular system so admirably adapted for attachment is wanting in

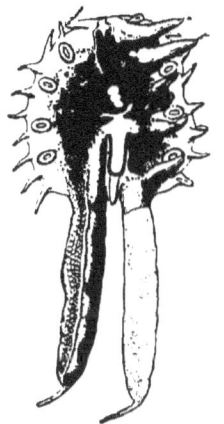 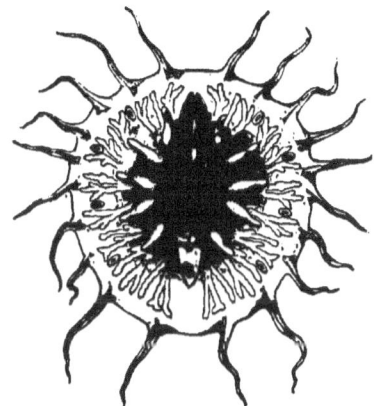

Fig. 419. — Myzostoma filicauda. ⅐. (Von Graff.)   Fig. 420.— Myzostoma Agassizii. ⅐. (Von Graff.)

the type which moves about freely on its host. In another group, a male and female inhabit a common cyst, caused by the presence of the parasite on the arm-joint or pinnule. *Myzostoma filicauda*, the host of *Antedon Hagenii*, is one of the species with caudal appendages (Fig. 419), while *Myzostoma Agassizii* represents a type with long filiform cirri. (Fig. 420.) Another group forming no cyst has only short cirri. The cysts are sometimes sausage-shaped and situated on the disk of the host, or, like the cyst formed by *M. cysticolum* (Figs. 421, 421 *a*) on the arms of *Actinometra meridionalis*, they resemble plant galls.

Fig. 421. Myzostoma cysticolum. ⁴⁄₁. Parasite of Actinometra meridionalis.   Fig. 421 *a*. (Von Graff.)

XX.

CHARACTERISTIC DEEP-SEA TYPES. — ACALEPHS.

CTENOPHORÆ AND HYDROMEDUSÆ.

As with fishes, a number of the deep-sea medusæ are occasionally taken at the surface, and undoubtedly many of the rarer of our jelly-fishes are deep-water forms which have accidentally found their way to the surface. To these probably belongs one of the most graceful of our jelly-fishes, *Ptychogena lactea* (Fig. 422), which swims at a considerable depth below the surface.

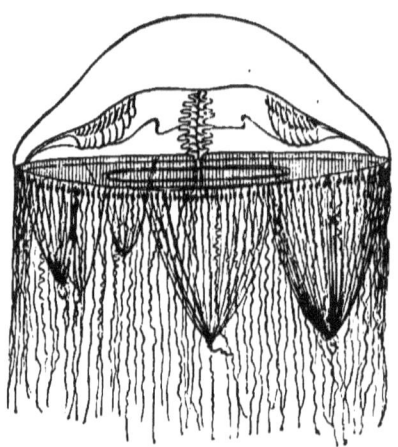

Fig. 422. — Ptychogena lactea. ¦.

The action of the light, and the increase of temperature at the surface, suffice to kill the animal in a short time. As soon as it reaches the surface, the disk loses its transparency, the genital organs become dull, and the medusa is soon completely decomposed, showing that the new conditions are totally unlike those under which it habitually thrives.

From the character of their development we may either find medusæ on the bottom in their fixed younger hydroid stages, or we may collect them alive from the surface in an older stage. Others again are always pelagic, swimming freely on the surface in all their stages of growth, while a limited number of the so-called deep-sea medusæ perhaps inhabit the intermediate depths far below the surface, moving from the bottom towards the surface, or even occasionally reaching it.

Although many of the characteristic surface jelly-fishes have been mentioned in the general sketch of the Pelagic Fauna and Flora, a few deserve a more extended notice in the systematic account of the group. Among the ctenophores I may mention a singular genus, Ocyroë, which has passed unnoticed for over fifty years, since its discovery in 1829. Unlike the other members of the group, it makes use of its large lateral lobes as flappers, and thus propels itself through the water with great rapidity. It is true that other ctenophores may, to a limited extent, guide their movements by the gentle undulation of the lateral lobes of the body, but their principal means of locomotion are the rows of locomotive flappers, or combs, from which the group derives its name. In Ocyroë the movement is produced by the development of muscular fibres on the inner surface of the lobes. Ocyroë is also noted for structural features of the highest interest. As has been observed by Dr. Fewkes,[1] it combines characters which exist in the two groups into which the ctenophores have been divided. It stands intermediate between the groups, with marked characteristics of each. It is the only instance of a ctenophore with lateral lobes not provided with tentacles. The spotted Ocyroë, *O. maculata* (Fig. 423), was noticed near St. Vincent; and a species without spots, probably a young form, *O. crystallina*, was found at the Tortugas.

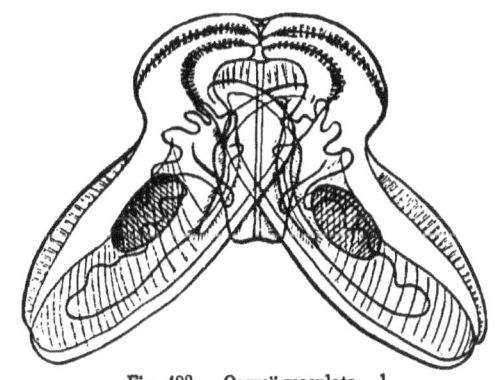

Fig. 423. — Ocyroë maculata. ⅓.

One of the largest and most stately genera of tentaculated

---

[1] Dr. Fewkes has prepared the greater number of the descriptions of acalephs here given.

ctenophores is the well-known *Eucharis multicornis* (Fig. 424), also found in the Mediterranean. This genus, which had before escaped observation on this side of the Atlantic, was observed at the Tortugas and at Key West.

Among the medusæ called Discophoræ by Agassiz, one of

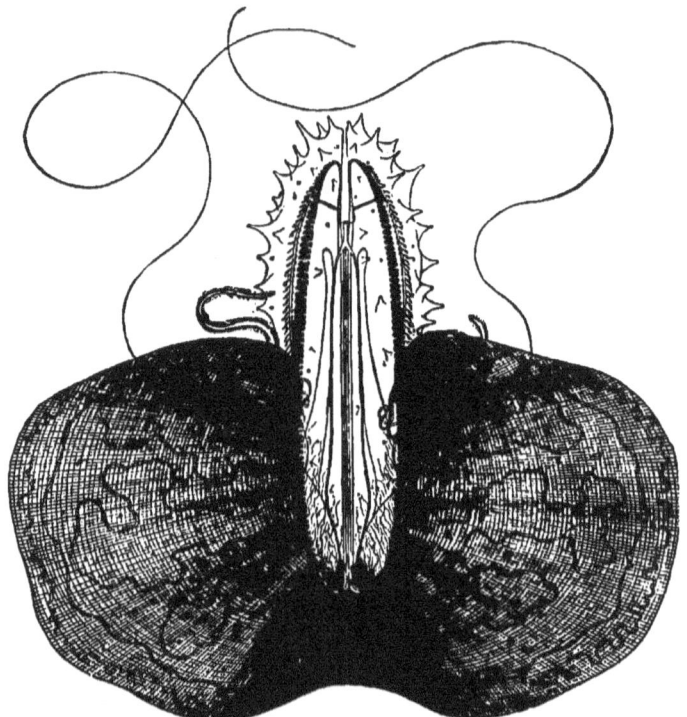

Fig. 424.—Eucharis multicornis. ½. (Chun.)

the most interesting forms is *Dodecabostrycha dubia* (Fig. 425), the largest specimen measuring no less than nine inches in height. Several specimens of a dark claret-color were brought up in the trawl, and it is very probable, from the systematic affinities of this medusa, that, like its allies, the Rhizostomæ, it lives on the bottom, rarely coming to the surface. Belonging also to the true deep-sea medusæ are Periphylla, Atolla, and a few allied genera. The first genus has a more or less pointed conical bell, widening below into a funnel-shaped margin, the upper and lower parts of the bell being divided into well-marked regions separated by a characteristic furrow. The margin is formed by a number of gelatinous blocks closely fitted together, which serve as supports for important organs called socles. These support tentacles, marginal sense bodies, and thin leaf-shaped lappets which have given the genus its name. The

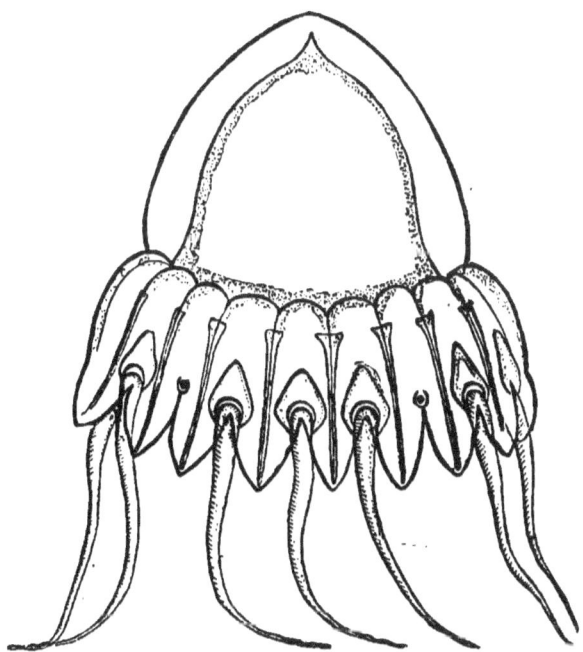

Fig. 425. — Dodecabostrycha dubia. ⅓.

stomach hangs down from the under side of the bell, and in its spacious receptacles are found prominent filaments. The color is blue. The American species *P. hyacinthina* (Fig. 426) extends as far north as the coast of Greenland.

None of these so-called deep-sea medusæ, however, present such remarkable features as the species of Atolla. The genus has thus far been taken by the "Challenger" in the Antarctic Ocean, on the borders of the South Atlantic and South Indian oceans, at the depth of about 2,000 fathoms. It is represented by a single species, *A. Wyvillei*. In the Gulf Stream and North Atlantic we have two species of Atolla, discovered by the " Albatross." They do not appear to be confined to deep water, but sometimes approach the surface. No discophore has as many sense segments as Atolla; and a marked feature of the oral surface of the bell is the large muscle found on the under

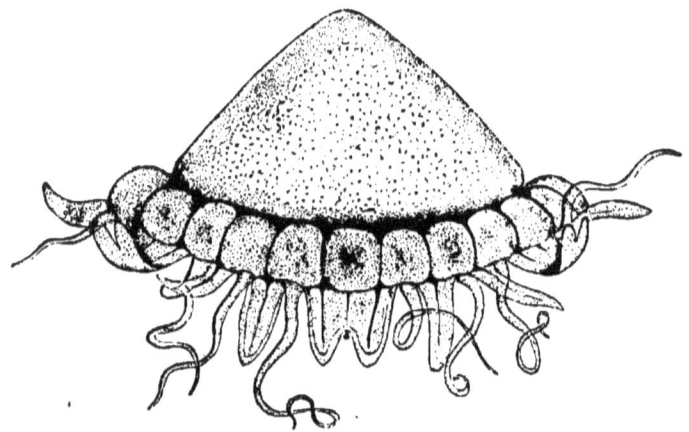

Fig. 426. — Periphylla hyacinthina. ¾. (Fewkes.)

side of the corona. The ovaries of Atolla consist of eight kidney-shaped bodies arranged about a large and spacious stomach, which assumes the form of an inflated bag, opening into a recess in the walls of the corona, from which canals extend into the tentacles and sense-bodies. *A. Bairdii* is here figured. (Fig. 427.)

Some of the most interesting medusæ discovered by the "Blake" belong to the Siphonophoræ. They are eminently pelagic in character, and wide-spread in their distribution. Previously to the "Blake" expeditions we knew only a few genera of these beautiful animals from the American coasts. Although genera of siphonophores occur in some of the most northern localities visited in Arctic exploration, the home of the group is essentially in the warmer waters. This group seems to be most varied and rich in the West Indian area. Before 1880,

CHARACTERISTIC DEEP-SEA TYPES. — ACALEPHS. 133

not more than five genera were known from the Western Atlantic, while at the present time that number is more than doubled.

Of the aberrant group of Rhizophysidæ no less than three species are now known from the Gulf Stream. One of the most

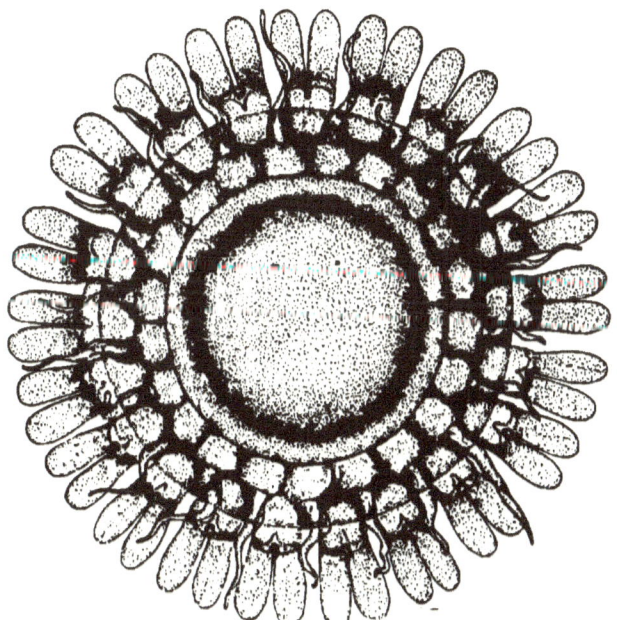

Fig. 427. — Atolla Bairdii. $\frac{3}{5}$. (Fewkes.)

characteristic species of the group, Pterophysa, has been mentioned in the chapter on the Pelagic Fauna.

Agalma Okenii (Fig. 428) is common in the Gulf Stream; it is easily recognized by the rigid nature of the colony, and by this can at once be distinguished from the Agalma found at Newport. The end of the axis opposite the float bears thick covering scales, while in the Newport Agalma the scale is leaf-like, and not cubical or polygonal.

One of the least known genera of Physophoræ is the genus Athorybia. It is remarkable in many ways, and differs from all known physophores in the character of its covering scales and the absence of nectocalyces, whose function is in part taken by the covering scales. They are capable of a slight motion on their attachment, and by this movement an active propulsion is produced. The float is large, and the stem very much reduced in length. The genus is interesting from its resemblance to a young

Fig. 429. — Gleba hippopus. ⅔. (Fewkes.)

Fig. 428. — Agalma Okenii. ¼. (Fewkes.)

stage of Agalma having no nectocalyces, in which a similar circle of covering scales is found. A new species, *A. formosa*, from the Florida Keys, has been added to the medusæ of the Gulf Stream.

The close resemblance of the swimming-bells of one genus of the floatless siphonophores to a horse's hoof suggested the name of *hippopus* to designate a wide-spread Mediterranean species (*Gleba hippopus*, Fig. 429) found in the Gulf Stream by the "Blake." In its affinities, Gleba is one of the most problematical of all the siphonophores. Like the physophores, it has two rows of nectocalyces, but no true float or covering scales. Moreover, in the physophores the nectocalyces nearest the float are the smallest and the last to form, while those at the opposite end are larger. In Gleba the bells at the anterior extremity are fully formed, while those at the posterior end are least developed.

We have two or three species of a distinct group of siphonophores, known as the Calycophoræ, one of the most common of which is *Diphyes acuminata*. (Fig. 430.) Another species, belonging to the genus Epibulia, was also collected; it is similar to a Mediterranean species, and is probably the same as that recorded from the coast of Greenland by Leuckart. The genus Abyla, *A. trigona*, was found in the Caribbean Sea, and fragments of a large Praya were observed near the Tortugas. I have already alluded to this group of siphonophores as driven into Narragansett Bay during the summer.

The first extensive report on deep-sea hydroids was based upon the collections made by Pourtalès in the Straits of Florida. They are described by Professor Allman, in one of the most important memoirs ever published on this group. The subsequent explorations of the "Blake" added a number of genera possessing most important morphological characters. As has subsequently been found in other collections of deep-sea hydroids, a majority of the genera collected belong to the Plumularidæ. A species of the genus Aglaophenia (*A. crenata*)

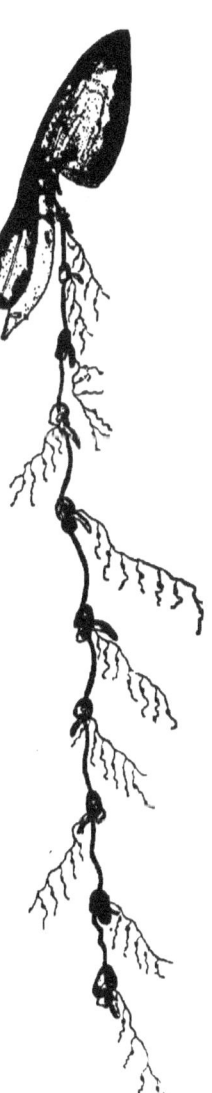

Fig. 430. — Diphyes acuminata. ¾. (Fewkes.)

was dredged from 1,240 fathoms, over 300 fathoms deeper than the greatest depth at which any plumularian was collected by the "Challenger." The Tubularians, so common in shallow water, do not seem to extend to any considerable depths. A characteristic plumularian is the stately *Aglaophenia bispinosa* (Fig. 431), dredged off Alligator and Tennessee reefs, from 200 fathoms, surpassed in size by very few hydroids. The corbulæ (Fig. 432) are very beautiful, and present a most instructive illustration of the morphology of the organ. The lower

Fig. 432.— Aglaophenia bispinosa, magnified. (Allman.)

part of the stem is composed of tubes, which, at rather regular intervals, become curiously contorted into knob-like projections. (Fig. 433.) They become separated at the extreme lower end, where they form a large entangled mass of filaments.

Cryptolaria conferta (Fig. 434), forming crowded entangled tufts, was dredged off Cojima, Cuba, in 450 fathoms. On the

Fig. 435. — Cryptolaria conferta, magnified. (Allman.)

Fig. 434. — Cryptolaria conferta. ¼. (Allman.)

branches of one of the specimens occurred here and there irregularly fusiform shaped bodies (Fig. 435), the nature of which

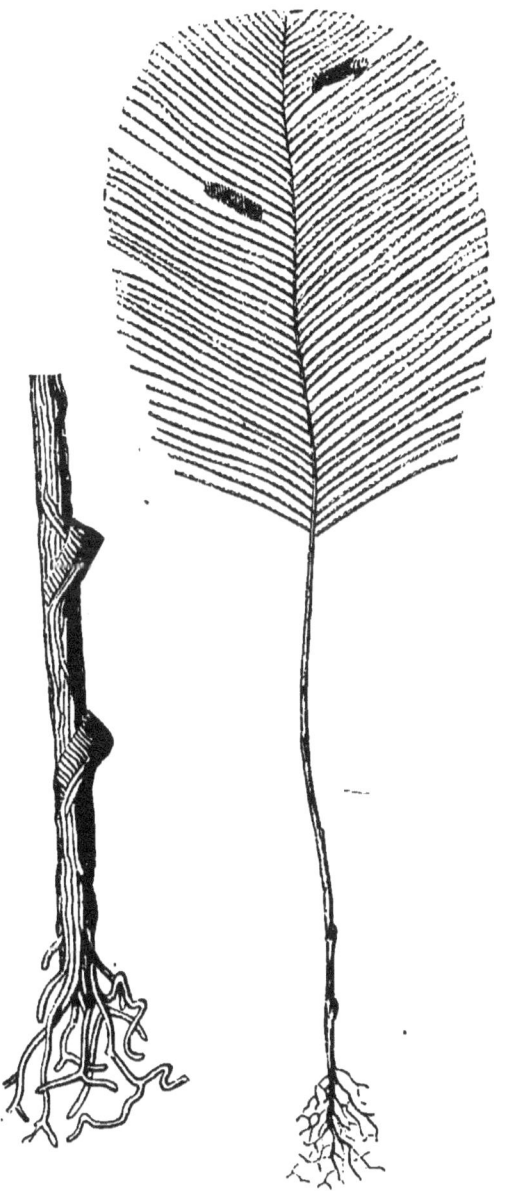

Fig. 433. — Lower part of stem of Fig. 431. (Allman.)   Fig. 431. — Aglaophenia bispinosa. ♀. (Allman.)

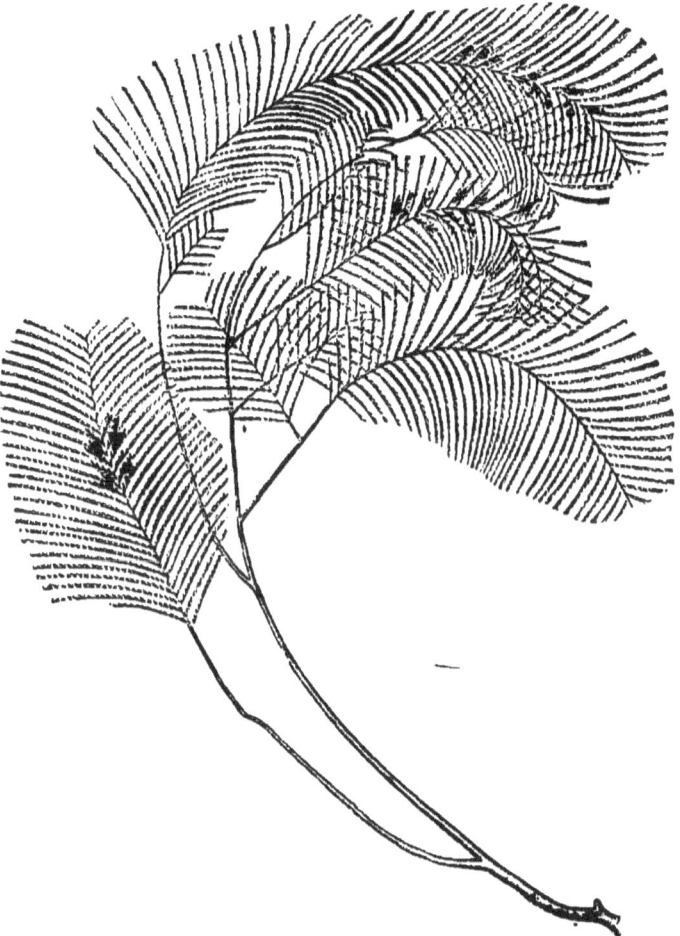

Fig. 430.— Cladocarpus paradisea. ½. (Allman.)

Fig. 437. — Hippurella annulata. ⅔. (Fewkes.)

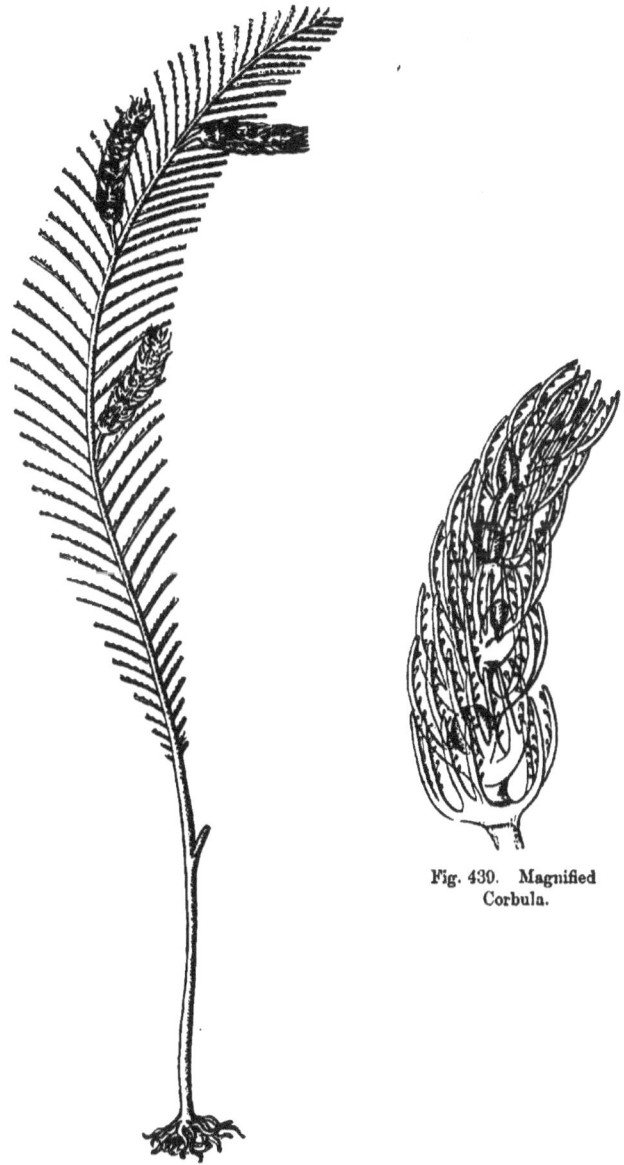

Fig. 439. Magnified Corbula.

Fig. 438. ¼.
Callicarpa gracilis. (Fewkes.)

is still problematical, surrounding the branch where they occur like minute sponges. They are found to consist of a multitude of flask-shaped receptacles.

The genus Cladocarpus was established by Allman for a remarkable plumularian obtained in the eastern part of the North Atlantic during one of the expeditions of the " Porcupine." *Cladocarpus paradisea* (Fig. 436), a beautiful species, very striking from its deep and widely separated hydrothecæ, was dredged off Tennessee Reef, and off the Samboes, from 174 fathoms.

Hippurella is a genus founded by Allman for hydroids in which the basal ends of the branches carry normal pinnæ, while the outer end of the same bear verticillately arranged ribs modified for sheltering the sexual bodies. *Hippurella annulata* grows in tufts, numerous undivided stems springing from a common base. (Fig. 437.) It is of a rather rigid habit; it was dredged off Pacific Reef, from 283 fathoms.

In Callicarpa we have whole branches specialized and modified for the protection of the sexual bodies. In *Callicarpa gracilis* (Fig. 438) the gonosome closely resembles a spike of wheat, and springs by a short peduncle immediately from the main stem. (Fig. 439.)

The most important of the family of Plumularidæ devoid of movable nematophores is Pleurocarpa, dredged from the neighborhood of the island of St. Vincent in 95 fathoms. In the single known specimen the gonosome (Fig. 440) certainly is the most extraordinary modification of the branch serving as a protection for the sexual bodies thus far found among plumularians. The basket-shaped structures called corbulæ, which serve the same purpose in other genera, are, as Allman has shown, modified pinnæ, and not, as in Hippurella, Callicarpa, and Pleurocarpa, a branch or portion of a branch bearing pinnæ modified to become specialized bodies with the form of corbulæ.

Fig. 440. Pleurocarpa ramosa; magnified. (Fewkes.)

HYDROCORALLINÆ.

To the hydroids we should add the account of the Hydrocorallinæ, which until recently were supposed to be true corals. Professor Agassiz, however, observed the animal of Millepora, and traced its acalephian affinity.

Fig. 441.—Animal of Millepora. ⁴⁴⁄₁. (Agassiz.)

The polyps of Millepora are most difficult to observe (Fig. 441), not only on account of their small size, but also from their extreme sensitiveness to contact with air. Agassiz's observations have been confirmed by several investigators, especially by Moseley, who has greatly increased our knowledge of the group, and has in addition shown that other families of corals, the Stylasteridæ and Helioporidæ, belong with the Milleporidæ to a natural group for which he has proposed the name Hydrocorallinæ. They are all characterized by having reproductive, prehensile, and digestive zoids composing the community (Fig. 442), reminding us thus somewhat of the siphonophores.

Fig. 442. — Millepora nodosa, Dactylozoid Gastrozoid; magnified. (Moseley.)

The best known member of the group is the shallow-water Millepora (Fig. 443), which is represented in deep water in the Caribbean and Florida districts by *Pliobothrus symmetricus*.

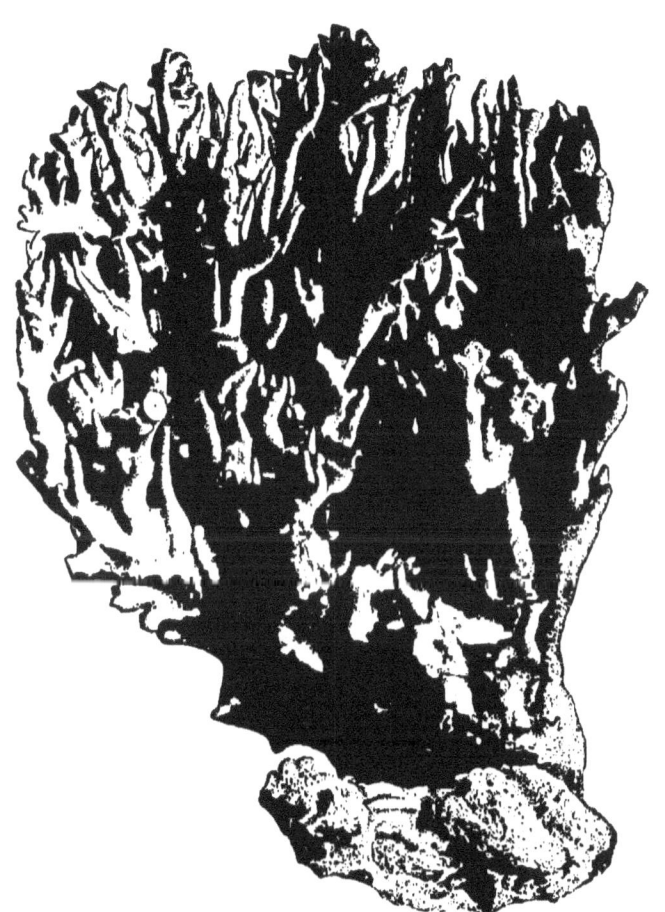

Fig. 443. — Millepora alcicornis. ⅔. (Agassiz.)

CHARACTERISTIC DEEP-SEA TYPES. — HYDROCORALLINÆ. 139

(Fig. 444.) It has also been found by the "Porcupine" expedition at from 500 to 600 fathoms, in the cold area to the northward of the British Islands.

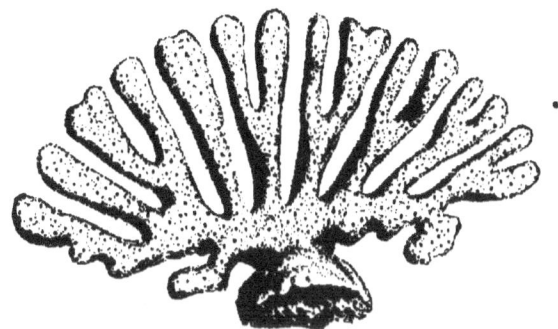

Fig. 444. — Pliobothrus symmetricus. ¼. (Pourtalès.)

The other family of the group added to the West Indian fauna by deep-sea dredgings is that of the Stylasteridæ; in some of the genera the simple circular digestive opening is drawn out into elongate chambers, while in some genera a

Fig. 445. ¼. Fig. 445 a. ¼.
Cryptohelia Peircei. (Pourtalès.)

tongue-like process or a lid covers in part or wholly this opening. Such an expansion of one edge of the calycle to form a lip (Fig. 445), folded over the opening, we find in *Cryptohelia Peircei*. (Fig. 445 a.)

Among the most beautiful and delicate of the milleporian corals found on rocky bottom is *Stylaster filogranus* (Fig.

446), of a light pink, fading into white in the younger branch-

Fig. 446. — Stylaster filogranus. ¼. (Pourtalès.)

lets. The color is diffused through the entire thickness of the corallum. Another common Stylaster is *Distichopora foliacea*

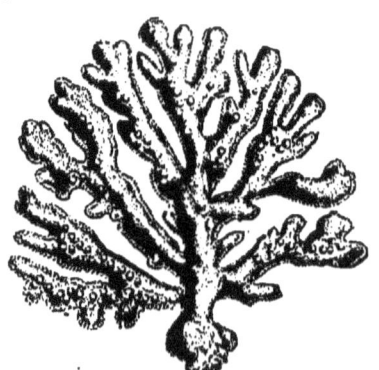

Fig. 447. — Distichopora foliacea. ¼. (Pourtalès.)

(Fig. 447), characterized by its small calycles, not placed in a furrow, irregular lateral pores, and serrated edge.

CHARACTERISTIC DEEP-SEA TYPES. — HYDROCORALLINÆ.

The most massive of our deep-sea corals is *Allopora miniacea*. (Fig. 448.)

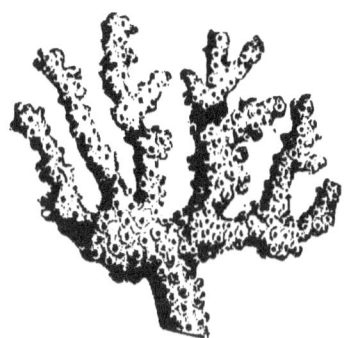

Fig. 448. — Allopora miniacea. ¼. (Pourtalès.)

XXI.

CHARACTERISTIC DEEP-SEA TYPES. — POLYPS.

HALCYONOIDS AND ACTINOIDS.[1]

AMONG the Anthozoa the deep-water groups of the West Indian district are most interesting. There are specimens of an Umbellula, a genus first accidentally brought up from deep water off the coast of Greenland early in the last century, and figured by Ellis. His specimens were lost, and Captain von Otter was the first to rediscover this interesting genus. The "Blake" dredged fine specimens of Umbellula in deep water in several localities in the West Indies. Our species of Umbellula appears to be *U. Güntheri* (Fig. 449), discovered by the "Challenger." A second species has since been found on our east coast by the Fish Commission.

A number of fine Pennatulæ were brilliantly phosphorescent, of a bluish tint. Their light is very strong, a single Pennatula lighting up a whole tub full of water. *Pennatula aculeata* (Fig. 450) is a common species off our coast, extending from Norway to the Banks of Newfoundland, and as far south as 33° north latitude. Of the peculiar club-shaped genus Kophobelemnon (Fig. 451) the "Blake" collected only a single specimen, but it has been dredged in considerable numbers by the Fish Commission. In certain localities it extends to a depth of over 2,000 fathoms.

Fig. 451. — Kophobelemnon scabrum. ¼. (Verrill.)

Several species of long sea-wands seem to be

[1] The account of these Anthozoa has been prepared from the reports of Professor Verrill on collections of the "Blake" and "Albatross."

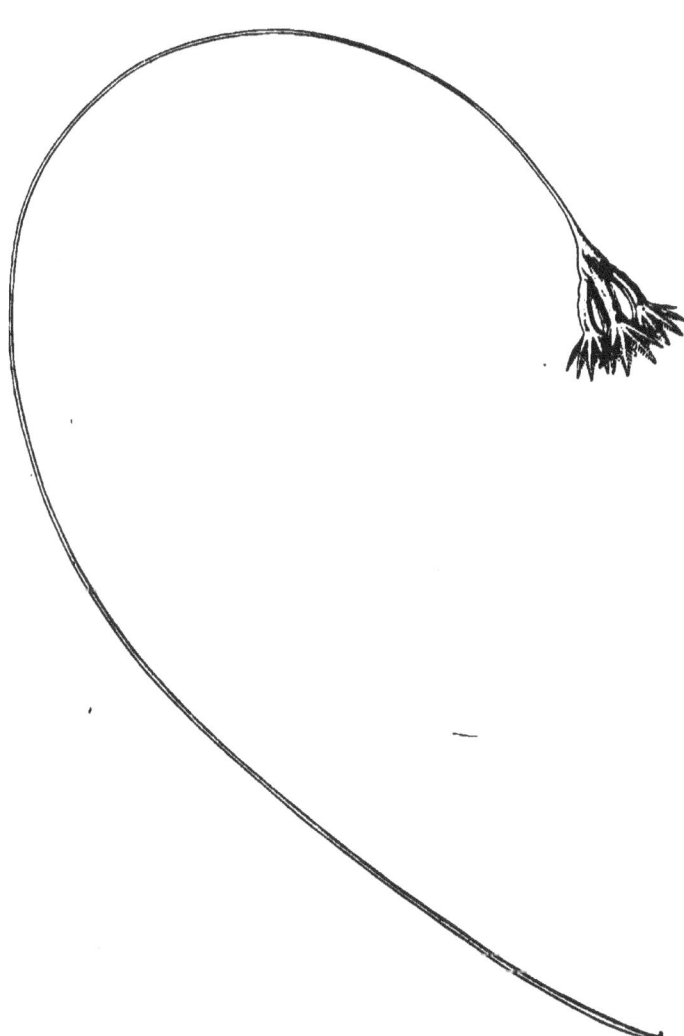

Fig. 440. — Umbellula Güntheri. $\frac{1}{4}$.

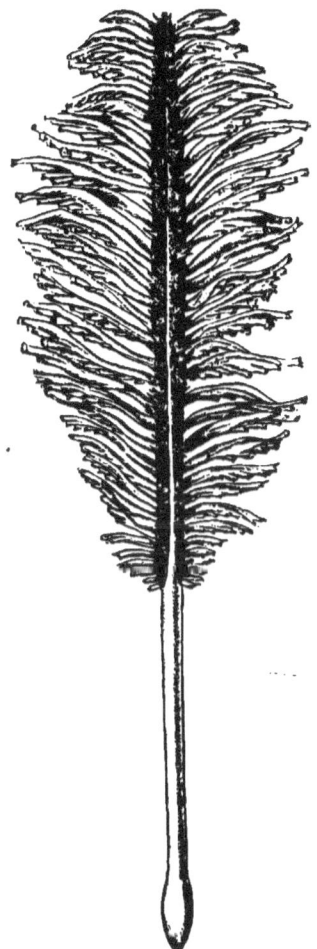

Fig. 450.—Pennatula aculenta $\frac{2}{7}$. (Koren & Danielssen.)

Fig. 452. — Anthoptilum Thomsoni. ½. (Kölliker.)

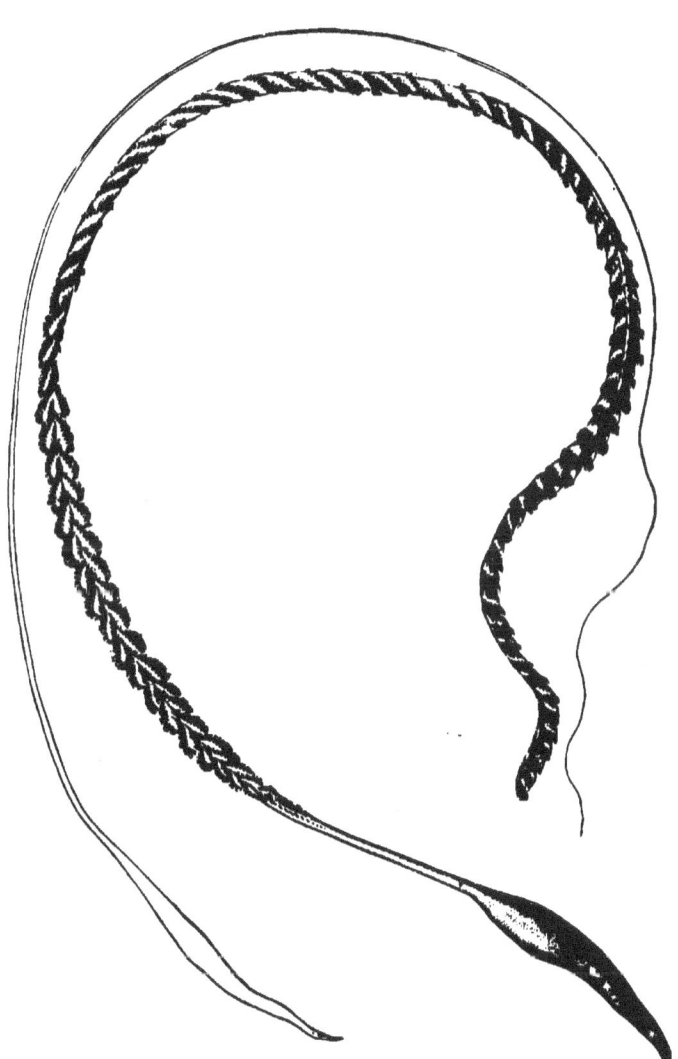

Fig. 453. — Balticina finmarchica. ½. (Koren & Danielssen.)
The axis is figured in outline.

the favorite abode of many kinds of ophiurans, and of sea-anemones, which are attached to the bare portions of the axis. We may mention among them a large species of Anthoptilum (Fig. 452), and a species of Balticina. (Fig. 453.) The extremity of the axis of many of these wands is frequently laid bare by injuries. These naked spaces, as has been observed by Professor Verrill, are nearly always occupied by a peculiar Actinia (Actinauge), of which the sides of the flat base spread out longitudinally so as to wrap around the axis of the polyp and meet on the opposite side, forming a regular sheath by the coalescence of opposite edges. (Fig. 454.) The base of adjoining Actiniæ coalesces in the same manner, and thus forms a continuous covering over the dead polyps.

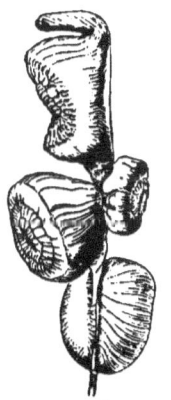

Professor Kölliker, who examined the "Challenger" collection of Pennatulæ, came to the conclusion that the deeper portions of the Pacific and Atlantic oceans contain very few Pennatulæ at a certain distance from shore, and that these appear to have a wide distribution along the shores; the higher groups especially being characteristic of shallower water, while the simpler forms, the representatives perhaps of an extinct fauna, inhabit the greater depths.

Fig. 454.— Actinauge nexilis. ⅔. (Verrill.)

The gorgonians are well represented in deep water by peculiar genera, of which the base is specially adapted for living in the mud, where it branches in all directions penetrating the soft ooze as if with roots; all the shallow-water species having usually a flat expansion of the base, by which they attach themselves to solid substances, rocks, mollusks, etc.

Many of these gorgonians are of an orange or reddish orange color; and the most characteristic of these is the elegant *Dasygorgia Agassizii* (Fig. 455), a plumose much-

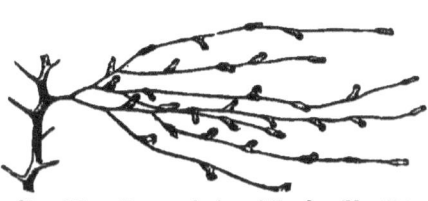

Fig. 455.— Dasygorgia Agassizii. ⅓. (Verrill.)

branching coral with slender terminal twigs, while the main branches are spirally arranged, with a slender brilliantly iridescent calcareous axis. The polyps are large, and placed rather far apart. It belongs to the deep-sea family of Chrysogorgidæ (Fig. 456), established by Verrill for such gorgonians as have usually an iridescent axis with spiral branches. They are among

Fig. 456. — Chrysogorgia. ½.

the most beautiful and interesting of the gorgonians. A unique and striking species of the group is Iridogorgia Pourtalesii (Fig. 456 a) with its regular upright spiral main stem, and long, flexible, undivided branches, arranged in a single row nearly at right angles to the axis; forming a broad spiral like the skeleton of a spiral staircase. The species are remarkable for their elegance of form, and for the brilliant lustre and iridescent colors of the axis, in some of a bright emerald-green, in others like burnished gold or mother-of-pearl. The known species are all

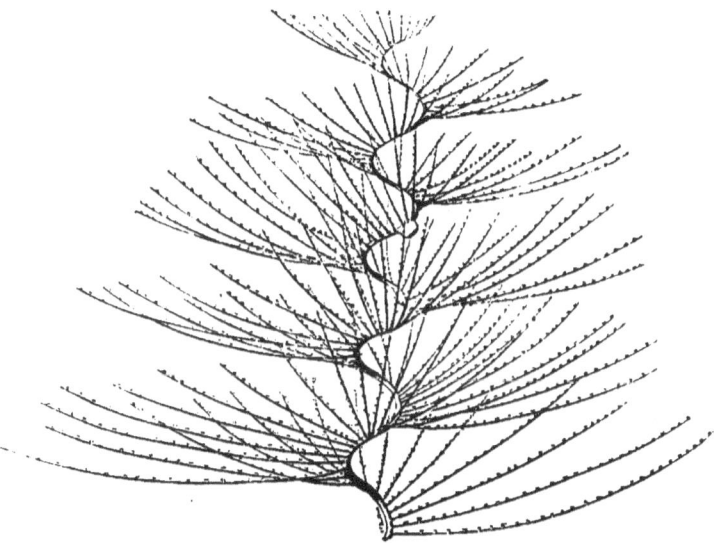

Fig. 456 a. — Iridogorgia Pourtalesii. ¼.

inhabitants of deep water, and with the exception of *Dasygorgia Agassizii*, which occurs off the New England coast, are all from the West Indies.

A large species is *Lepidogorgia gracilis*, which grows to a height of nearly three feet. A smaller gorgonian, but perhaps the most common off our east coast, extending from 200 to about 1,300 fathoms, is *Acanella Normani* (Fig. 457), a branching bush-like orange-brown coral. It grows to a height of about a foot, and is nearly as broad as high, its branches growing out three or four together from the joints.

Ceratoisis ornata is a large and beautiful species peculiarly characteristic of deep water in all latitudes, its golden or bronzy chitinous joints contrasting finely with the clear ivory-white calcareous ones. Lepidisis is a gorgonian growing in the shape of a tall thin stem a yard or more in height, its axis divided into joints alternately long and short; the longer ones white, hollow, and calcareous, and the shorter ones horny brown. We should also mention *Primnoa Pourtalesii* (Fig. 458), a plumose gorgonian with regularly pinnate branchlets all in one plane. To this genus belongs also the huge bush coral Primnoa,

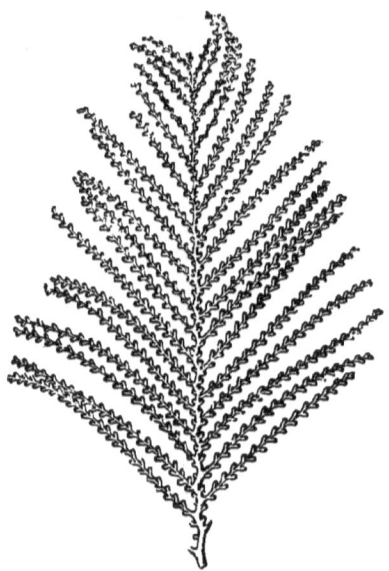

Fig. 458. — Primnoa Pourtalesii. ¾. (Verrill.)

which grows to the height of man, and has an axis as thick as a man's leg.

Many of the gorgonians are beautifully phosphorescent when brought to the surface, and their closely clustered branches, as in Calyptrophora (Fig. 459) are the abode of hosts of crustacea, annelids, mollusks, and echinoderms, which find shelter there from their enemies.

The Actinidæ, or sea-anemones, so common in shallow water, are represented by a number of species in our deep waters; many of them are finely colored, some of them developing a peculiar base adapted to soft bottoms, representing perhaps, as has been suggested by Verrill, a primitive type from which the

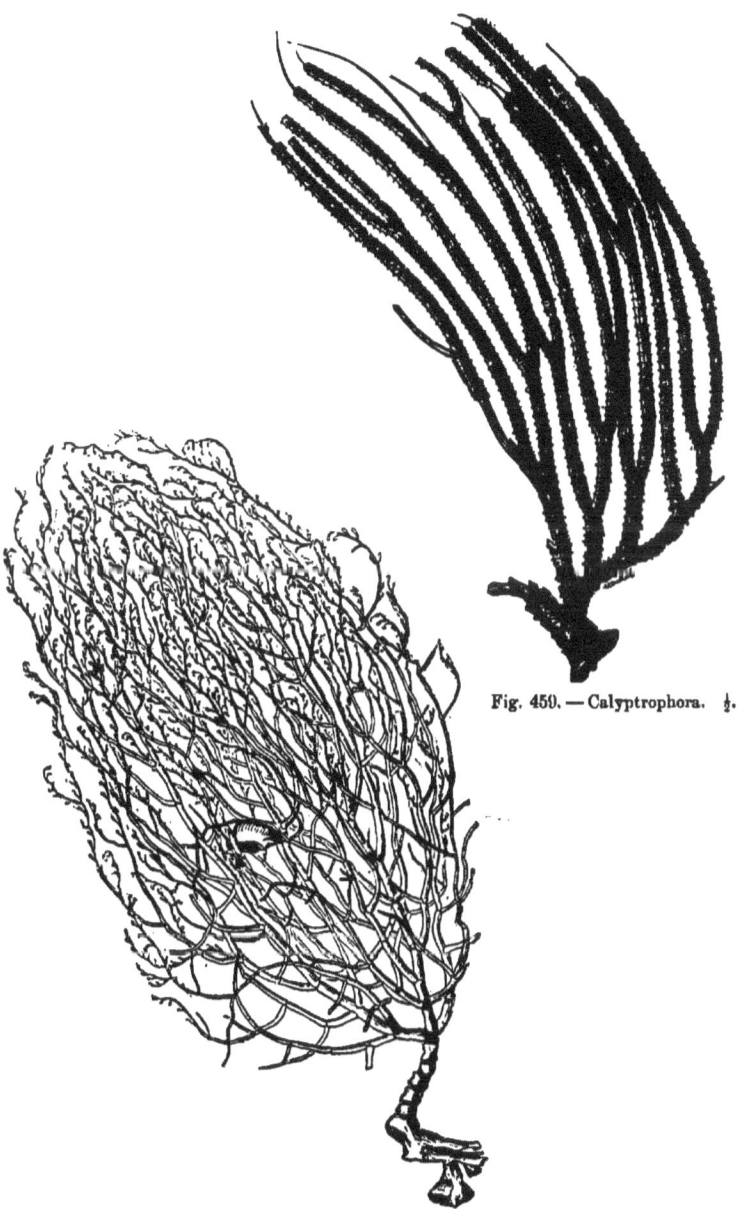

Fig. 459. — Calyptrophora. ½.

Fig. 457. — Acanella Normani. ½.

CHARACTERISTIC DEEP-SEA TYPES. — POLYPS. 147

few surviving Pennatulidæ may have been derived. But owing to the difficulty of determining satisfactorily animals of this family from alcoholic specimens, we shall notice only a few species which have been figured from life by Verrill.

Sagartia abyssicola (Fig. 460) is often found attached to the tubes of Hyalinœcia. A large red or orange species of Actinauge is *A. nodosa* (Fig. 461), the column of which is covered with hard warts arranged in rather regular transverse and vertical rows, diminishing in size from the top of the column towards

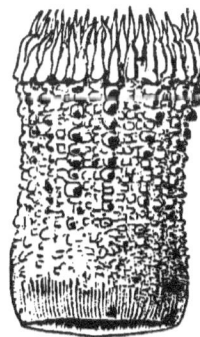

Fig. 461. — Actinauge nodosa. ½. (Verrill.)

the base. Specimens of four inches in diameter and six inches in height are often brought up in the dredge. It has been dredged off our eastern coast, and extends from the Grand Banks to Cape Hatteras. Its bathymetrical range is from 50 to 600 fathoms. From the tentacles and upper part of the column is secreted

Fig. 460. — Sagartia abyssicola. ⅔. (Verrill.)

an abundant mucus, which is highly phosphorescent. As has been suggested by Verrill, these Actiniæ, anchored as they are in the mud by a basal bulb, probably lose their power of locomotion gradually with their development, and finally when adult remain fixed, although they certainly move freely about when young, like other shallow-water actiniæ.

Epizoanthus belongs to a group of actiniæ usually forming irregularly shaped incrusting masses and incapable of locomotion. The polyps have a thick leathery column of a bluish or grayish-brown color. Two species are quite common along the east coast of the United States, in depths varying from 75 to 600 fathoms. (See Fig. 235.)

CORALS.[1]

A series of fine specimens of *Caryophyllia communis* (Fig. 462) well shows their mode of growth. The young is erect, with a thin peduncle attached to a small pebble or shell ; as it

Fig. 462. Fig. 462 a.
Caryophyllia communis. ¼. (Pourtalès.)

grows in height, the support not being sufficient, it falls over on its broadest side, and, growing upward to keep the calycle above the mud, the curved base is produced. (Fig. 462 a.)

Stenocyathus vermiformis is a very elongate coral resembling an annelid tube. Specimens frequently occur having a living and growing polyp at either end. (Fig. 463.) These specimens are generally somewhat curved, as if they had been lying in the mud with both ends turned up and projecting.

Fig. 463. — Stenocyathus vermiformis. ¼. (Pourtalès.)

[1] The account of the corals here given is taken from the various reports of Pourtalès on the "Blake" collections.

Two species of the genus Thecocyathus have been dredged, and are not uncommon in from 100 to 315 fathoms. One of these, *T. cylindraceus*, is here figured. (Fig. 464.) The genus is interesting as dating back to the lias; it is not known from any of the formations intermediate between the lias and our epoch. The recent forms present, therefore, a comparatively rare instance of the reappearance of a genus apparently extinct through a considerable succession of ages.

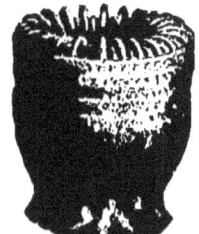

Fig. 464. — Thecocyathus cylindraceus. $\frac{3}{1}$. (Pourtalès.)

Deltocyathus italicus (Figs. 465, 465 a–d) is an exceedingly

Fig. 465 a.

Fig. 465 c.

Fig. 465.

Fig. 465 d.

Fig. 465 b.

Deltocyathus italicus. $\frac{3}{1}$. (Pourtalès.)

variable living form of a tertiary fossil common in Sicily. The polyp of a large living specimen, dredged in 115 fathoms off the Tortugas, was whitish, with short club-shaped tentacles. A most variable species is *Paracyathus confertus*. (Fig. 466.) *Stephanotrochus diadema* (Fig. 467) seems to be a characteristic deep-sea type. It has been dredged in 734 fathoms off

Guadeloupe, and in 1,200 fathoms fine living specimens of *Fla-*

Fig. 466. — Paracyathus confertus. ¼.
(Pourtalès.)

Fig. 467. — Stephanotrochus diadema. ¼.
(Pourtalès.)

bellum Moseleyi (Figs. 468, 468 *a*) were obtained; this species has an extensive geographical range. Its ally, *Flabellum*

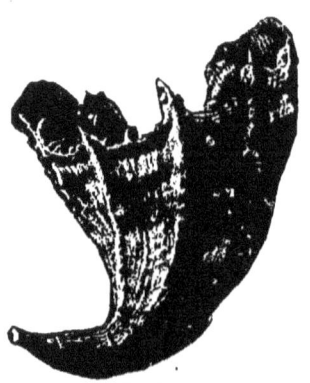

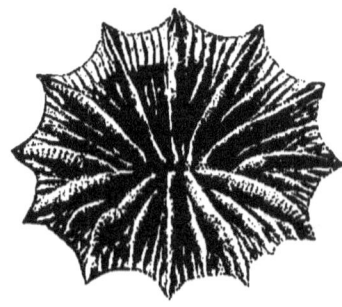

Fig. 468.   Fig. 468 *a*.
Flabellum Moseleyi. ¼. (Pourtalès.)

Goodei, is quite common off the east coast of the United States, and grows to a considerable size. The stout tentacles and disk are of a salmon-color.

Desmophyllum Riisei (Fig. 469) is a species growing in clusters; it ranges from 88 to 120 fathoms off Montserrat, Dominica, and Martinique. *Desmophyllum solidum* (Fig. 470) is the West Indian representative of several species of this type from the tertiary beds of Sicily. A very fine specimen of *Des-*

mophyllum crista-galli came up attached to the stem of a Prim-

Fig. 469. — Desmophyllum Riisei. ½. (Pourtalès.)

Fig. 470. — Desmophyllum solidum. ⅔. (Pourtalès.)

noa. *Rhizotrochus fragilis* (Fig. 471) was obtained in about

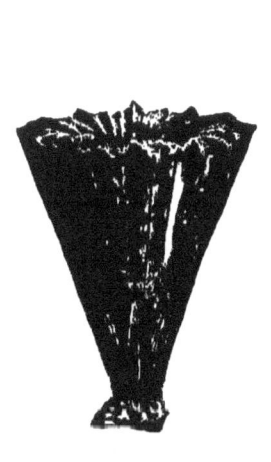

Fig. 471. — Rhizotrochus fragilis. ⅔. (Pourtalès.)

Fig. 472. — Lophohelia prolifera. ½. (Pourtalès.)

forty different casts along the Florida Reef, in depths varying from 49 to 324 fathoms. It was most abundant between 100 and 200 fathoms. The color of the polyp is greenish or pale brick-red. *Lophohelia prolifera* (Fig. 472) has a very extensive distribution in the Atlantic, and is a common Caribbean species.

One of the most elegant of the West Indian corals is the pure white *Amphihelia rostrata* (Fig. 473), which must have spread

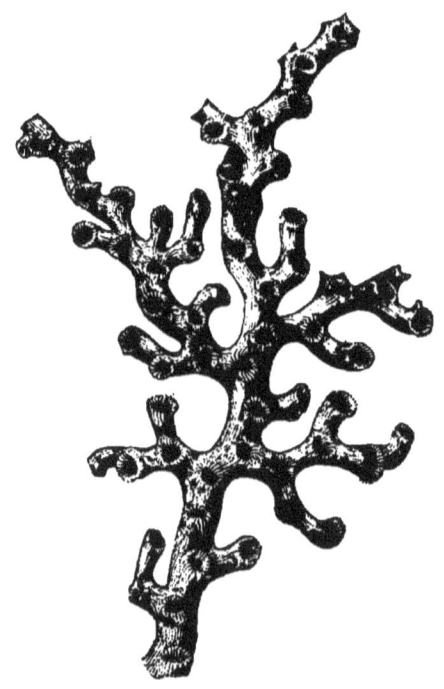

Fig. 473. — Amphihelia rostrata. ½.

at least twelve centimetres. It has been dredged to a depth of nearly 900 fathoms. *Axohelia mirabilis* (Fig. 474) is very common in the Caribbean, and is rather variable. Many specimens are deformed by barnacles occupying the end of the branches, which soon become entirely covered by the coral, leaving only a small opening. As representatives of one of the most natural of the families of corals, we may mention *Thecopsammia socialis* (Fig. 475) and *T. tintinnabulum*, of which the living polyp is of a handsome pinkish orange color.

The Fungidæ are a very characteristic shallow-water form in the Pacific, and it is interesting to note that from deep water

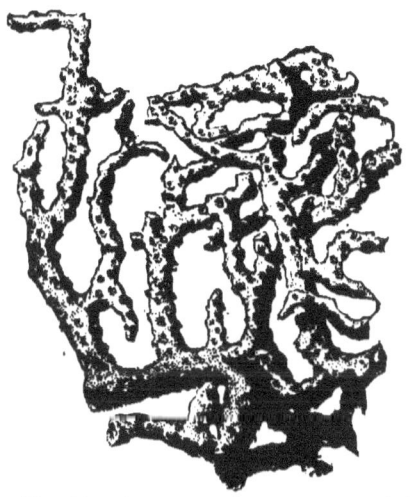
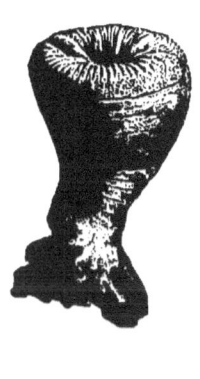

Fig. 474. — Axohelia mirabilis. ½. (Pourtalès.)   Fig. 475. — Thecopsammia socialis. $\frac{1}{1}$,$\frac{6}{1}$. (Pourtalès.)

the dredge has brought up three small, simple species, the first simple Fungia found in our seas. Of these we may mention *Fungia symmetrica* (Fig. 476), found by the dredgings of the "Challenger" to be one of the most common deep-sea corals. It has a world-wide distribution; it occurs in the West and South Atlantic, and in the North and South Pacific, and has a very extended bathymetrical range; it has been dredged by the "Challenger" in 30 fathoms and in 2,900 fathoms, and in all intermediate depths. The range of temperature which it sustains varies from 1° to 20° C. It has been found by the "Blake" ranging from 175 to 800 fathoms in the Gulf of Mexico, the Straits of Florida, and the Carib-

Fig. 476. — Fungia symmetrica. ?. . (Pourtalès.)

Fig. 477. — Dinseris crispa. ⅔. (Pourtalès.)

bean. The specimens of *Fungia symmetrica*, and *Diaseris crispa* (Fig. 477), if found in a sea where larger Fungiæ were common, would naturally be considered as the young of one of them.

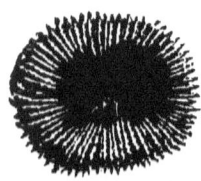

Antillia explanata (Fig. 478) is the first species of this tertiary genus found living.

Fig. 478. — Antillia explanata. ⅔. (Pourtalès.)

Attached to the test of an Asthenosoma, from a depth of 373 fathoms near Montserrat, came up a fine specimen of the delicate *Leptonemus discus* (Fig. 479), in which the corallum is reduced to a mere lacework.

Fig. 479. — Leptonemus discus. ⁹⁄₇. (Challenger.)

Among corals recalling extinct types are specially to be mentioned Haplophyllia (Fig. 480), Duncania, and Guynia, which were surmised by Pourtalès to belong to the Rugosa, an order established by Milne-Edwards and Haime for a large number of fossil corals, abundant in palæozoic times. Their chief characteristic is the development of the septa from four primary ones (Fig. 481), whilst in all of the living corals the primary number is six. In addition, the chambers are closed by floors. Ludwig has shown, however, that this tetrameral arrangement of the

Fig. 480. Haplophyllia paradoxa. ¼. (Pourtalès.)

Rugosa is only apparent, there being originally six primary septa, two of the systems remaining generally undeveloped. The polyp of *Haplophyllia paradoxa* is scarlet, with about sixteen rather long tentacles. In another species, *Duncania barbadensis*, the polyp is deep flesh-colored, and there are from 25 to 30 conical tentacles with inflated tips.

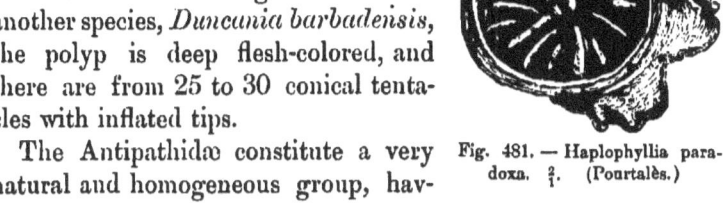

Fig. 481. — Haplophyllia paradoxa. ⅔. (Pourtalès.)

The Antipathidæ constitute a very natural and homogeneous group, having the property of secreting a horny polypidom. One of the most common West Indian species is *Antipathes spiralis;* it

Fig. 482. — Antipathes spiralis. ⅒. (Pourtalès.)

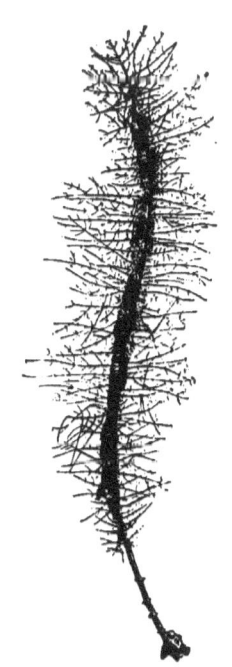

Fig. 483. — Antipathes columnaris. ¾. (Pourtalès.)

has been dredged from no less than twenty-three stations, in depths ranging from 45 to nearly 900 fathoms. The polyps of this species are alternately large and small, with very long digitiform tentacles. The figure (Fig. 482) represents them as they are frequently disposed, the larger polyps alone being visible, while the smaller ones can only be seen in the profile view. At other times the tentacles are very much shortened and stiffened, and stand out from the axis. The singular mode of growth of *Antipathes columnaris* (Fig. 483) deserves a few words of description. The central hollow column is occupied by an annelid which appears to compel the corallum to form an abnormal growth of that shape. Every one of the specimens dredged was similarly affected, and the annelid was still in place in most cases. A similar action of parasitic annelids has been noticed in some true corals, such as Lophohelia, Stylaster, Allopora, and others.

XXII.

CHARACTERISTIC DEEP-SEA TYPES. — RHIZOPODS.

THERE must be, all over the bottom on which reticularian rhizopods have been found, thousands of undiscovered minute protozoans which have no solid tests. On account of the difficulty of examining on the spot the samples of bottom as they are brought up, we can only conjecture the physiology of these lowest types, which will undoubtedly be discovered whenever the proper methods for examination are employed. In the mean time, we must be satisfied with a knowledge of the types which have become known to us from their tests; but even these do not explain the structure of their animals; this is known to us only by comparison with that of their shallow-water allies.

No special report of the "Blake" Foraminifera has as yet been completed, but I am fortunate in being able to extract from the admirable memoirs of Brady on the "Challenger" Foraminifera, and of Dr. Goës on the Rhizopoda of the Caribbean, descriptions and figures of the principal types collected by us. Dr. Goës, during a stay of several years at St. Bartholomew, explored a considerable area with the dredge, to a depth of 400 fathoms, and, owing to the existence of extensive sunken plateaux and steep sloping banks, where the temperature falls rapidly, he was able to collect the majority of the types which we subsequently brought together from deeper waters, but which extend upwards to depths of 200 fathoms, or 150 even, and perhaps less.

Of the rhizopods the siliceous radiolarians play an unimportant part in the bottom deposits of the district explored by the "Blake." A few surface species were collected in the track of the Gulf Stream. Yet, judging from the well-known radiolarian earth of Barbados, there was a period when radiolarian ooze must have been an important deposit of the West Indian region, probably during the time when the Caribbean was connected with the Pacific.

The arenaceous types of foraminifera, on the contrary, abound in the bottom deposits of the Caribbean and Mexican districts, and along the Western Atlantic, and the principal families are all well represented in the "Blake" collections. On some bottoms, the rhizopods vie in the variety of their development with those found in some of the celebrated tertiary and cretaceous localities.

There is a marked absence of siliceous sand and a scarcity of siliceous spicules from the coralline and calcareous ooze, so that rhizopodan types are preëminently calcareous; only a few succeed in making up their tests entirely of siliceous particles. We shall therefore find associated siliceous and calcareous forms greatly differing in outward shape, but Dr. Goës is inclined to consider this as of small importance, and due entirely to the difference of materials employed by one and the same type, according to the character of the bottom, and that a sort of isomorphism is established between species formerly considered as belonging to either the arenaceous or vitreous groups.

Where there are such enormous changes going on during the growth of a species, it is natural that in this group, as well as in sponges, we should find it extremely difficult to retain our old notions of species; and until the careful investigations of Williamson, Parker, Carpenter, and Brady among the foraminifera, and of Haeckel among the sponges, but little systematic order had been established in these groups. Endless generic and specific names followed in rapid succession, till the task of identifying any form of these groups seemed hopeless.

While among the more highly organized invertebrates the effect of the nature of the bottom is seen rather in an association of animals characteristic of rocky, gravelly, muddy, or sandy districts, we find that in such groups as the sponges and rhizopods the nature of the bottom is an all-essential factor in modifying the organism.

The bottom of the slópes and plateaux, and of the area where rhizopods flourish, between 150 and 400 fathoms, consists mainly of a chalky, tough, amorphous ooze, — a modified pteropod and globigerina ooze. Mixed with this are grains of similar material, but of a greater consistency, together with dead shells of pelagic mollusks and foraminifers and a great number of the tests of dead rhizopods, which once lived on the bottom and among which flourished in great abundance the innumerable large and small species characteristic of the Caribbean district. The majority of the largest rhizopods occur on the bottom, which is covered with the coarser fragments of corallines, annelid tubes, and other pieces of limestone, soldered together more or less compactly, and transformed into rough masses and lumps resembling coarse mortar or gravel.

Associated with the arenaceous, siliceous, and calcareous rhizopods which undoubtedly live upon the bottom, we find the tests of Globigerinæ, Hastigerinæ, Pulvinulinæ, and many others which have also been observed as pelagic. For a time it was supposed that the deposits so widely extended were due to Globigerinæ living on the bottom, but the evidence gradually brought forward by Bailey, Johannes Müller, Pourtalès, Major Owen, and especially by Mr. Murray of the "Challenger," seems to leave no doubt that the Foraminifera to which the globigerina ooze is due are pelagic, the ooze being formed by the dead shells after they have reached the bottom.

One of the most common types of rhizopods is *Biloculina ringens* (Figs. 484, 484 a, 484 b), a most abundant form in

Fig. 484. $\frac{9}{1}$.

Fig. 484 a. $\frac{9}{1}$.
Biloculina ringens. (Goës.)

Fig. 484 b. $\frac{10}{1}$.

deep water in the Atlantic; it is found nearly everywhere, from the littoral region to a depth of 3,000 fathoms. Along our coast off Block Island, and in a portion of the area between

Norway, Bear Island, and Spitzbergen, *Biloculina ringens* forms the most important organic constituent of the bottom deposits, and Pourtalès and Sars have named this the Biloculina clay; but this term is hardly to be understood in the same sense in which we speak of globigerina ooze, the Biloculinæ forming but a very small proportion of the ooze deposit. (Fig. 485.)

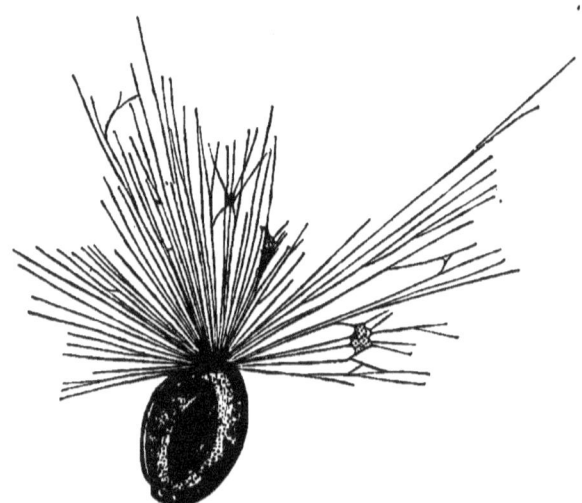

Fig. 485. — Biloculina tenera. With expanded pseudopodia. ♀. (Schultze.)

Orbiculina adunca (Fig. 486) is a very common deep-water form; it attains a diameter of 6 mm. In both Orbiculina and

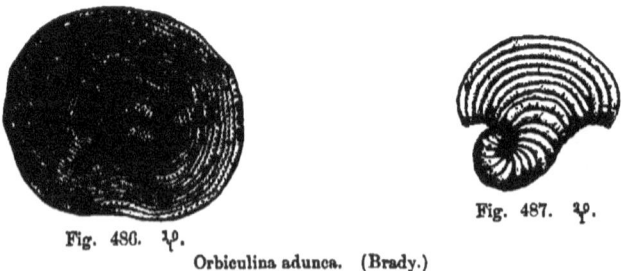

Fig. 486. ♀.
Orbiculina adunca. (Brady.)

Fig. 487. ♀.

its ally, Orbitolites, the young of the disk-like foraminifer is nautiloid (Fig. 487); but as the chambers of the adult increase in number they become more circular, and finally conceal the original nautiloid structure of the test. The genus dates back to the miocene. Except along the American coast, where the genus appears to be a deep-sea type, Orbitolites is found in shallow water; it is quite common on coral reefs. *Cornuspira foliacea* (Fig. 488), though it occurs in the arctic seas in great abundance in comparatively shallow water, is not uncommon in the pteropod ooze of the Caribbean.

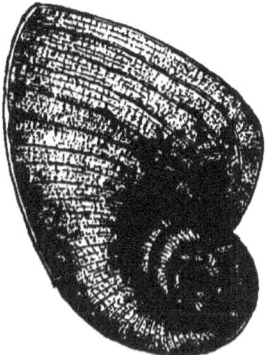

Fig. 488. — Cornuspira foliacea. ⅟. (Goës.)

Astrorhiza (Fig. 489) is a soft-tubed type remarkable for the absence of any definite aperture, the pseudopodia possibly finding their way out between the loosely aggregated sand-

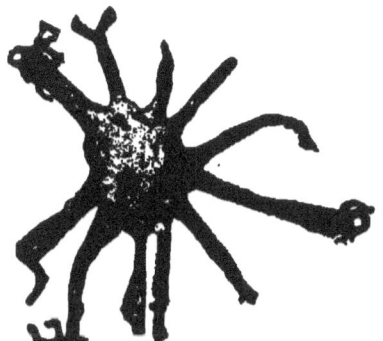

Fig. 489. — Astrorhiza limicola. ⅟. (Brady.)

grains of which it is composed. It has been found off Block Island and along the eastern coast of the United States, at moderate depths. The great variety in the composition and consistency of the test seems due in part to the material of the bottom, and in part perhaps to the great stillness of the waters in which it lives. This type was first described by Dr. Sundahl, in 1847, from specimens found in shallow muddy water on the Scandinavian coast. Allied to Astrorhiza is a not uncommon species of Pelosina, a flask-shaped rhizopod with thick walls composed of globigerina ooze. Another type of the same group resembles a long-necked flask with branching calcareous projections, but thickly covered outside with fragments of sponge spicules. Large fragments of a mammillary mass of loosely

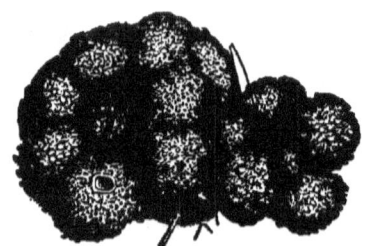

Fig. 490. — Sorosphaera confusa. ¹⁰⁄₁. (Brady.)

crumbling character, *Sorosphæra confusa* (Fig. 490), are not uncommon from the Caribbean globigerina ooze.

Hyperammina elongata (Figs. 491, 491 a), almost a cosmo-

Fig. 491. Fig. 491 a.
Hyperammina elongata. ⅔. (Goës.)

politan species, builds a long, slender tube which attains a length of 15 mm. Sometimes it is constructed of siliceous sand and sponge spicules. It has been dredged in the North Atlantic to nearly 1,800 fathoms. The genus dates back to the silurian. *Rhabdammina abyssorum* (Fig. 492), composed entirely of siliceous sand-grains, is one of the most characteristic forms of the deeper-water rhizopods of the Caribbean, Gulf of

Fig. 492. — Rhabdammina abyssorum. ⅔. (Brady.)

CHARACTERISTIC DEEP-SEA TYPES. — RHIZOPODS.   163

Mexico, and Gulf Stream. The three (Fig. 493) or four armed varieties often come up in great quantities in the dredge, and attain a length of from 16 to 20 mm. *R. abyssorum* has a worldwide distribution; it was discovered by the elder Sars, and de-

Fig. 493. — Rhabdammina abyssorum. ⅔. (Brady.)

scribed in his first list of animals living in deep water off Norway. This species presents many interesting modifications dependent on external conditions, and its polymorphism seems remarkable; it is triradiate, quadriradiate, or a straight tube, including all their possible combinations. A small straight form of the genus, *R. linearis* (Fig. 494), is also frequent near the 500-fathom line.

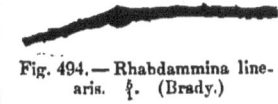

Fig. 494. — Rhabdammina linearis. ⅔. (Brady.)

One of the species of Lituolinæ, *Reophax scorpiurus*, attains a length of 10 mm. It builds its test loosely of siliceous sand

Fig. 495.   Fig. 495 a.
Reophax scorpiurus.   ⅔. (Goës.)

and sponge spicules. (Figs. 495, 495 a.) A widely spread form crowded with nipple-shaped protuberances, *Thurammina papil-*

lata (Fig. 496), is frequently brought up in the globigerina ooze from depths greater than 400 fathoms. The group dates back to the jurassic, and seems to be a characteristic deep-sea type in all the oceanic basins. *Ammodiscus tenuis* (Fig. 497), taken by the "Challenger" off New York in 1,300 fathoms, is a recent representative of a very common palæozoic type of the carboniferous period. According to Brady, Cyclammina (Fig. 498) represents in our seas the highest type of arenaceous foraminifers. The genus is characterized by the labyrinthian structure of the test (Fig. 499), and is abundant in depths below 100 fathoms in the West Indian region.

Fig. 496. — Thurammina papillata. ⁴⁄₁. (Brady.)

Fig. 497. — Ammodiscus tenuis. ¹⁄₁. (Brady.)

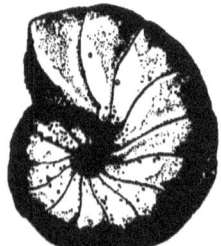

Fig. 498. ⁸⁄₁.
Cyclammina cancellata. (Brady.)

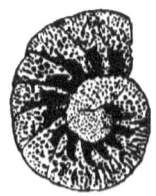

Fig. 499. ⁴⁄₁.

Most variable in the shape and structure of their shells are the Textularinæ. A very common type of the group is the cosmopolitan *Textularia sagittula* (Fig. 500), which attains a length of 6 mm.; it has been dredged in the Atlantic in 2,675 fathoms. Another abundant form, which dates back to the cretaceous, is the compact and thick-walled *T. trochus* (Figs.

Fig. 500. —Textularia sagittula. ¹⁄₁. (Goës.)

501, 501 a), with its smooth surface covered with fine pores.

 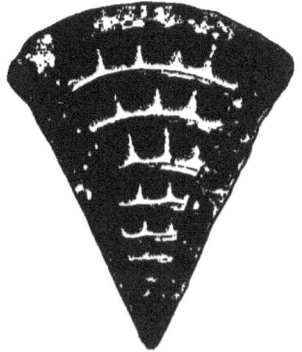

Fig. 501.   Textularia trochus. ⅜. (Goës.)   Fig. 501 a.

Valvulina triangularis (Figs. 502, 502 a), a North Atlantic

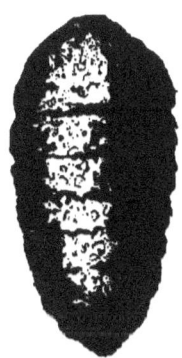 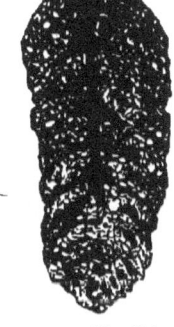

Fig. 502.   Valvulina triangularis. ⅜. (Goës.)   Fig. 502 a.

foraminifer, is also one of the typical West Indian rhizopods; it is characterized by its loosely constructed test.

The Lagenidæ are a most widely distributed type; according to Brady, they are found in all seas, at all depths from the littoral region to 3,000 fathoms. One of the common species, *Lagena distoma*, is here figured. (Fig. 503.)

Fig. 503. — Lagena distoma. ²⁹⁰. (Brady.)

One of the most variable foraminifers is *Nodosaria radicula* (Fig. 504), an Atlantic species of wide distribution. It is known by innumerable specific names, and the list of its varieties, as given by Dr. Goës, fills no less than ten quarto pages, these varieties representing all those possible combinations of smoothness, roughness, and striation of the test, or in the shape of the chambers, which seemed important to their describers. In many other species, also, names have been multiplied indefinitely. A species widely spread, both over the coralline bottom and ooze, is *Nodosaria communis* (Fig. 505), which attains a size of 22 mm. It closely resembles one of the cretaceous species, and dates back to the permian. From the same bottom comes the diminutive *Cristellaria crepidula* (Fig. 506), remarkable for its beautiful pearly shell. The West Indian specimens of *Cristellaria calcar* (Fig. 507) fully equal in size those from the chalk and tertiaries.

Fig. 504. Nodosaria radicula. $\tfrac{2}{1}^0$. (Goës.)

Fig. 505. Nodosaria communis. $\tfrac{1}{1}$. (Goës.)

Fig. 506.—Cristellaria crepidula. $\tfrac{2}{1}^0$. (Goës.)

Closely allied to the Nodosarinæ is *Sagrina dimorpha* (Figs. 508, 508 a), abundant in the ooze. It attains

Fig. 507. Cristellaria calcar. $\tfrac{1}{1}^0$. (Goës.)

Fig. 508. Sagrina dimorpha. $\tfrac{1}{1}$.

Fig. 508 a. (Goës.)

Fig. 509. Polymorphina ovata. $\tfrac{4}{1}^0$. (Brady.)

a size of 4 mm. in length. Living specimens of *Polymorphina ovata* (Fig. 509) have been obtained by the "Blake" and "Challenger" in the Caribbean district.

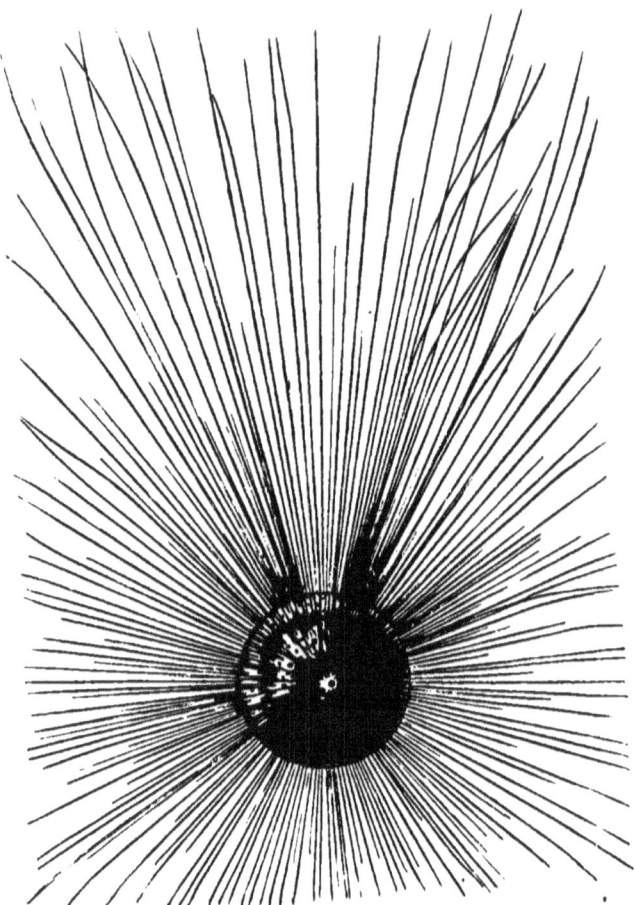

Fig. 510. — Orbulina universa. $\frac{50}{1}$. Surface. (Challenger.)

Orbulina universa, a cosmopolitan species, dates back to the lias and is very common in the tertiary. It is widely distributed on the coralline and ooze of the Caribbean, and one of the pelagic types most frequently found. Pourtalès discovered that bottom specimens of *O. universa* did not always consist of a simple chamber, but generally included three or four chambers (Fig. 510),.resembling young Globigerinæ more or less developed, and attached to the inside by slender spicules. Krohn observed the same in living specimens. It seems probable that the Globigerinæ in the chamber are resorbed, and that the visible spherical chamber is the last segment, considered at one time to be a special reproductive chamber, and capable of widespread existence. The Globigerinæ are eminently pelagic, some of the genera exclusively so, and the shells when alive are thin and transparent.

The shell of Globigerina is composed of a series of hyaline and perforated chambers of a spheroidal form, arranged in a spiral manner, with the apertures of each chamber opening round the umbilicus. The young shells are made up of fewer and comparatively larger chambers. The tests of *Globigerina*

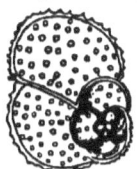
Fig. 511.

Fig. 511 *a*.
Globigerina bulloides. $\frac{1}{1}^6$. (Goës.)

Fig. 511 *b*.

bulloides (Figs. 511, 511 *a*, 511 *b*) and of *Orbulina universa* (Fig. 512) are among the most common deep-water rhizopods. Globigerinæ are spinous in their early stages, and probably more or less so when the shell has attained its full development, but the spines are of such extreme tenuity that, when taken with the tow-net, they are invariably broken. Bottom specimens have no spines, and these may be present perhaps only in the pelagic stage ; the delicate calcareous spines, from four to five times the diameter of the

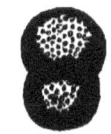
Fig. 512. — Orbulina universa. $\frac{1}{1}^6$. (Goës.)

shells, enabling them to float with greater facility by increasing their surface immensely. When alive, "the sheaves of these spines cross in different directions, and have a very beautiful effect." The inner chambers are filled with a colored sarcode, either red or orange. No trace of pseudopodia has as yet been observed, or any extension of the sarcode beyond the shell.

Globigerina bulloides has been found pelagic everywhere in the West Indies, as well as in the bottom dredgings of the Caribbean and the Gulf Stream. It is not so abundant after passing north of Cape Hatteras. I have not found it pelagic off the coast of the Middle States. Hastigerina is eminently a pelagic type. It had been known from the coast of South America many years previously to its rediscovery by the "Challenger." It is not an uncommon pelagic type off the Tortugas, and was found on one occasion, on a very calm day, swarming on the surface with *Globigerina bulloides*.

A minute scale-like foraminifer, *Discorbina orbicularis*, is commonly found in the coral reefs of the West Indies. Another peculiar form, also found living in the West Indian reefs,

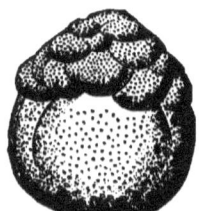

Fig. 513.—Cymbalopora bulloides. ⁴⁄₁. (Challenger.)

is Cymbalopora; one of the species of the genus, however, *C. bulloides* (Fig. 513), is also pelagic.

The most protean of West Indian rhizopods is perhaps *Carpenteria balaniformis*. (Fig. 514.) Its regular structure is

Fig. 514.—Carpenteria balaniformis. ⁴⁄₁. (Goës.)

rotaline, but, owing to its propensity for developing additional chambers from the upper extremity and from the chamber

CHARACTERISTIC DEEP-SEA TYPES. — RHIZOPODS. 169

walls, the greatest variety of forms and of deviation from the parent type results.

Pulvinulina auricula (Figs. 515, 515 a) is a handsome hya-

Fig. 515. Fig. 515 a. Fig. 516. Fig. 517.
Pulvinulina auricula. ¹⁄₁. (Goës.) Pulvinulina Menardii. ¹⁄₁. (Goës.)

line species, and its ally, *P. Menardii* (Figs. 516, 517), is one of the most common deep-water species. It is also pelagic.

Another deep-water form is *Truncatulina Ungeriana*. (Fig. 518.) The little *Polytrema miniaceum* (Fig. 519) is a delicate red parasitic foraminifer, occurring everywhere in the West Indies, which resembles certain minute corals. It has a long geological history, dating back to the devonian. Of the Nummulinidæ, *Polystomella crispa* is one of the most abundant West Indian types of moderate depths.

Fig. 518. — Truncatulina Ungeriana. ¹⁄₁. (Goës.)

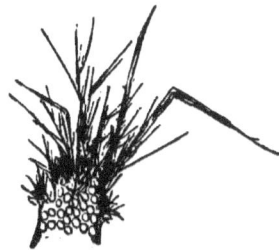

Fig. 519. — Polytrema miniaceum. ¹⁄₁

XXIII.

CHARACTERISTIC DEEP-SEA TYPES. — SPONGES.[1]

WE are led by the study of the Sponges to some of the most interesting biological problems. All our ordinary notions of individuality, of colonies, and of species are completely upset. It seems as if in the sponges we had a mass in which the different parts might be considered as organs capable in themselves of a certain amount of independence, yet subject to a general subordination, so that, according to Haeckel and Schmidt, we are dealing neither with individuals nor colonies in the ordinary sense of the words.

As Schmidt well says: "From the variability of all characters our ideas of an organism as a limited or centralized individual disappear in the sponges, and in place of an individual or a colony we find an organic mass differentiating into organs, while the body, which feeds itself, and propagates, is neither an individual nor a colony."

We shall specially dwell on the more prominent Hexactinellidæ and Lithistidæ of the Caribbean district. These groups date back to the lower silurian, and take an extraordinary development in the Jura; they are quite abundant in the upper cretaceous, but poorly represented during the tertiaries. Wyville Thomson was perhaps the first to insist upon the relationship of the Hexactinellidæ with types of former geological periods, the Ventriculites of the chalk. They, like the Lithistidæ, the remains of a second fossil family, are in decided minority in the seas of to-day.

The absence of siliceous and other sponges in the collections made along the northern part of the east coast of the United States is very striking, and although the number of specimens

[1] The account of the sponges has been prepared from the memoirs of Professor Oscar Schmidt on the Atlantic and Caribbean sponges.

of certain species was often very great, yet the continental fauna of that region is poor when compared with the wealth of species found in the Caribbean Sea and Gulf of Mexico.

Among the Hexactinellidæ one of the most common types is

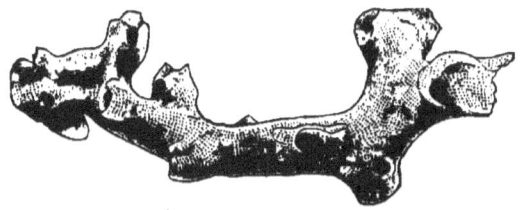

Fig. 520. — Farrea facunda. ⅔.

the variable *Farrea facunda* (Fig. 520), which occurs either as a simple or a somewhat complicated form; it is found at depths

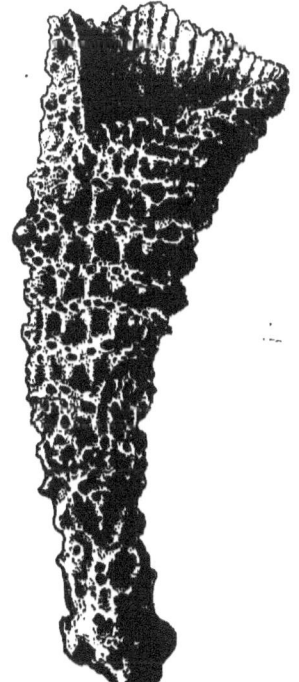

Fig. 521. — Lefroyella decora. ⅔.

of 300 to 1,000 fathoms. One of the finest, figured by Thomson, a species called *Lefroyella decora* (Fig. 521), was dredged by him from a depth of 1,075 fathoms off the Bermudas. It must have attained at least a foot in height.

Another most common and at the same time most exquisite type of Hexactinellidæ is *Aphrocallistes Bocagei* (Fig. 522),

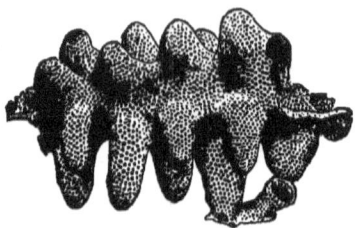

Fig. 522. — Aphrocallistes Bocagei. ⅓.

which has been dredged by the "Blake" in depths of from 164 to 400 fathoms. It is also found in the eastern basin of the North Atlantic. The network appears to be formed by the coalescence of stellate spicules. These sponges are often attached to corals and soldered together, so as to form large convoluted masses. Dactylocalyx is one of the most characteristic of the Caribbean types. The shape of *Dactylocalyx pumiceus* (Fig. 523) varies from that of a cup to that of a flat dish attached by a short stem. The surface is furrowed and

Fig. 523. — Dactylocalyx pumiceus. ½.

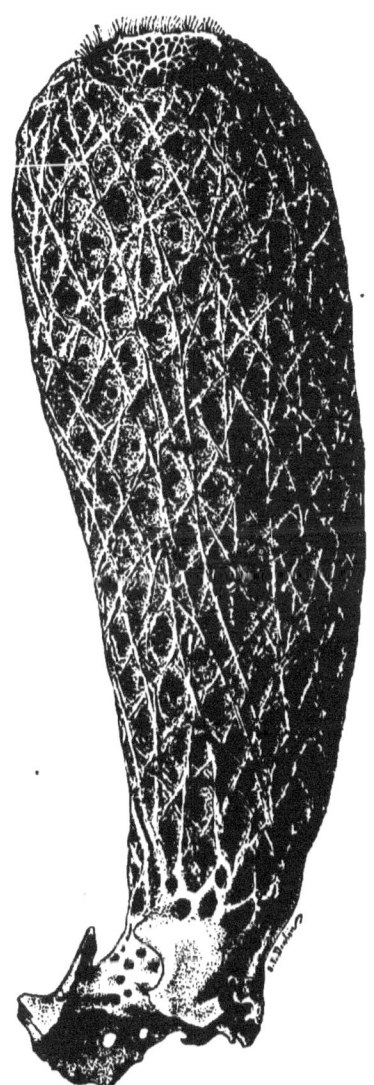

Fig. 524. — Regadella phœnix.

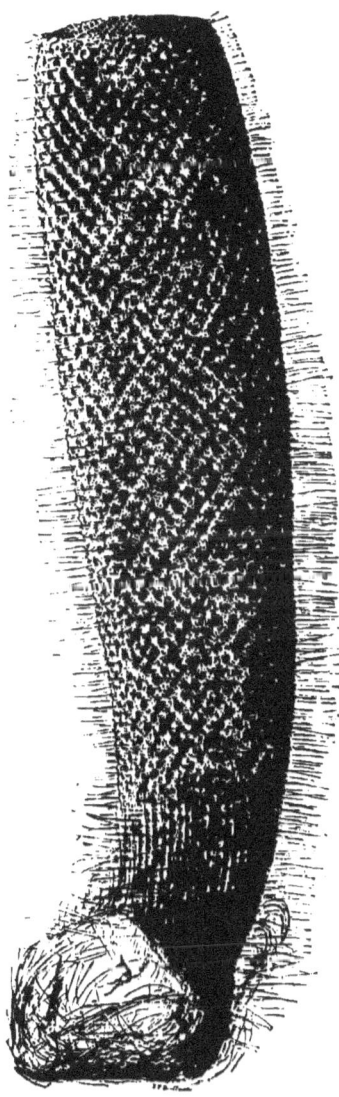

Fig. 525. — Euplectella Jovis. ½.

CHARACTERISTIC DEEP-SEA TYPES. — SPONGES. 173

perforated, and the sculpture is arranged radially with some degree of regularity.

The Euplectellidæ (known principally as Venus's Basket, from the Philippine Islands) are represented in the West Indian region by huge species, and by peculiar types adapted to a rocky bottom, such as *Regadella phœnix* (Fig. 524), while the typical Euplectellæ seem to have flourished best in ooze. *Euplectella Jovis* (Fig. 525) must have been at least 48 centimetres in length.

Hyalonema Sieboldii (Fig. 526), a cosmopolitan species, was

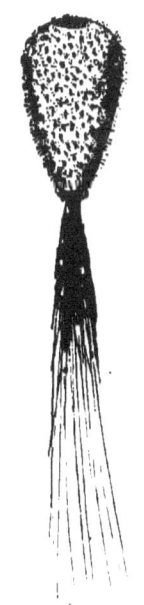

Fig. 526. — Hyalonema Sieboldii. ⅔.

also found near Grenada in 416 fathoms. The Japanese long deceived naturalists regarding a species of Hyalonema representing the bundle of siliceous spicules as the axis of a Gorgonia-like animal. (Fig. 527.) Leidy

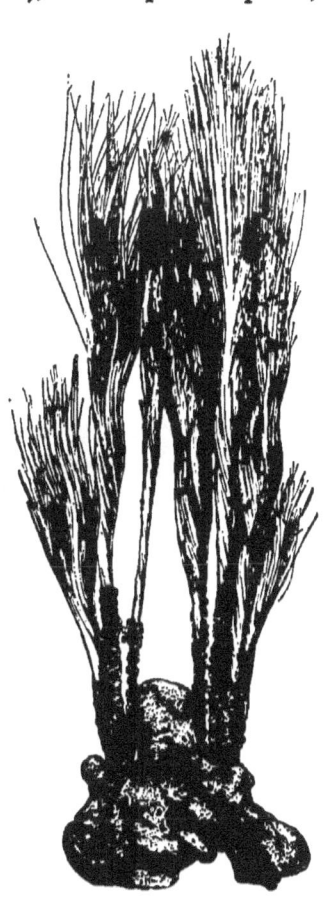

Fig. 527. — Spicules of Japanese Hyalonema with encrusting polyps. ½.

was the first to show that the sponge and siliceous cable were one organism, and the polyps mere parasites attached to it above the mud and below the sponge (Fig. 528), a view which has been fully confirmed. *Asconema setubalense*, a magnificent siliceous sponge, first dredged by Kent off the coast of Portugal, has a wide geographical distribution. Very fine specimens were collected by the "Talisman," and one of the adjoining figures

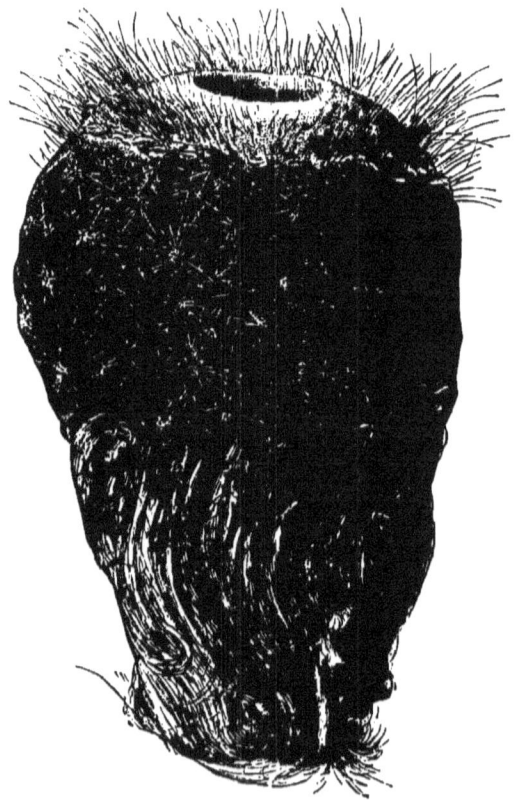

Fig. 531. — Holtenia Pourtalesii. ⅔.

(Fig. 529) is taken from one of the best preserved specimens of the French expedition. It is a common species in the West Indies, in from 300 to 600 fathoms. *Pheronema Annæ* (Fig. 530), first described by Leidy, is represented by some most

Fig. 528.—Japanese Hyalonema. ⅓.

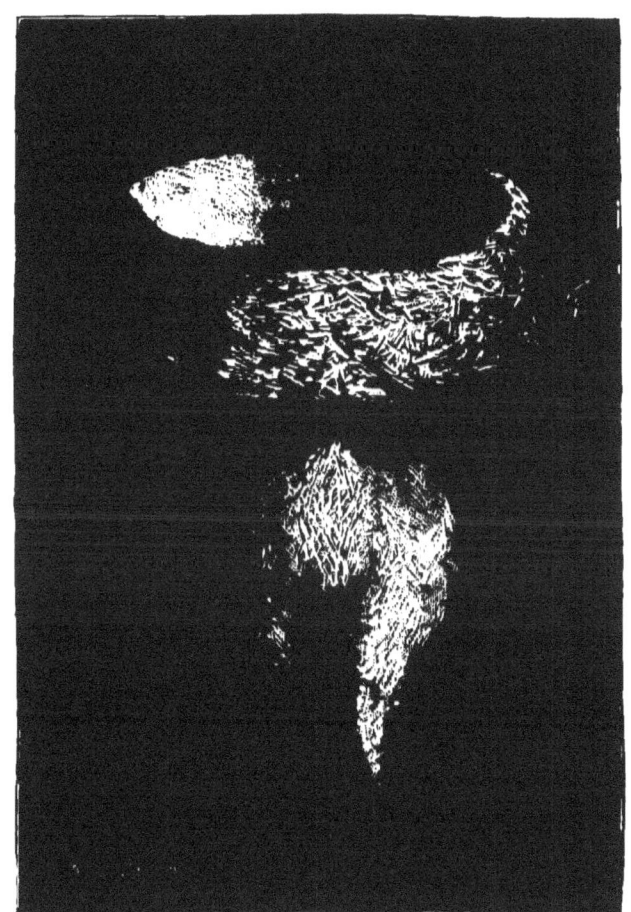

Fig. 520. — Asconema setubalense. ¼. (Filhol "Talisman" Ex.)

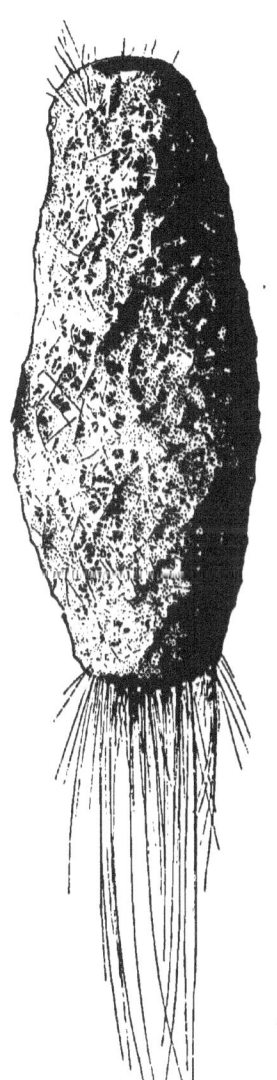

Fig. 530. — Pheronema Annæ. $\frac{2}{3}$.

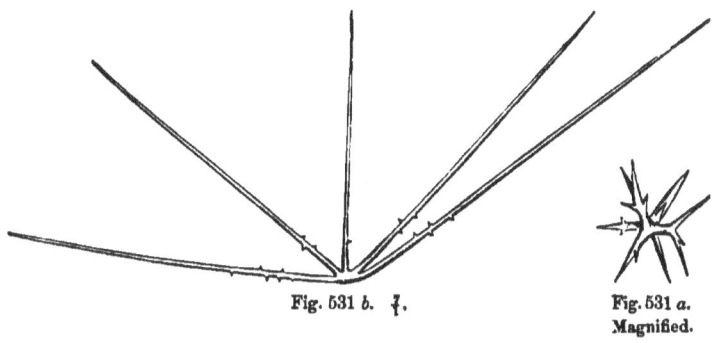

Fig. 531 b. ⁴⁄₁.   Fig. 531 a. Magnified.

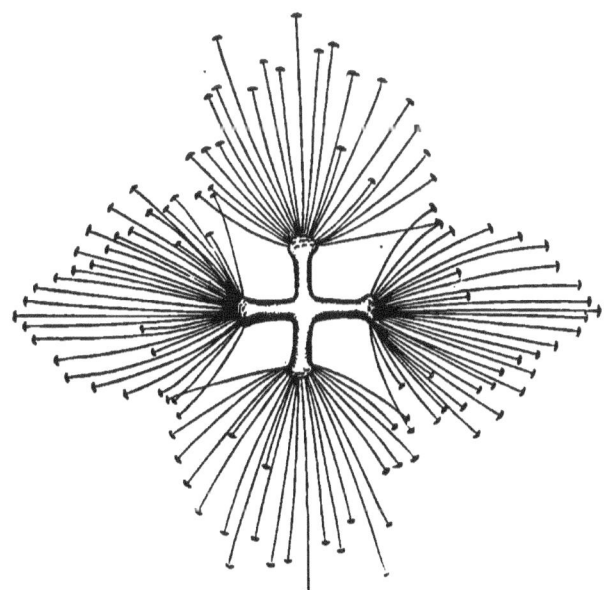

Fig. 531 c. Greatly magnified.
Figs. 531 a, 531 b, 531 c. Spicules of Holtenia Pourtalesii. (Schmidt.)

CHARACTERISTIC DEEP-SEA TYPES. — SPONGES. 175

beautiful specimens collected off Frederichstæd in 180 to 208 fathoms, in thick globigerina ooze. A fine *Holtenia Pourtalesii* (Figs. 531, 531 *a*, 531 *b*, 531 *c*) was collected by Pourtalès off Sand Key, in depths varying from 184 to 324 fathoms.

The group of Lithistidæ, as defined by Zittel, includes sponges, formerly united with the Hexactinellidæ, characterized

Fig. 532. — Vetulina stalactites. Greatly magnified. (Schmidt.)

by their connected calcareous spicules (Fig. 532), not built upon the three-axis type, but forming an apparently irregular maze.

The majority of the specimens of *Vetulina stalactites* (Fig. 533) are thick, undulating sheets, closely perforated with irreg-

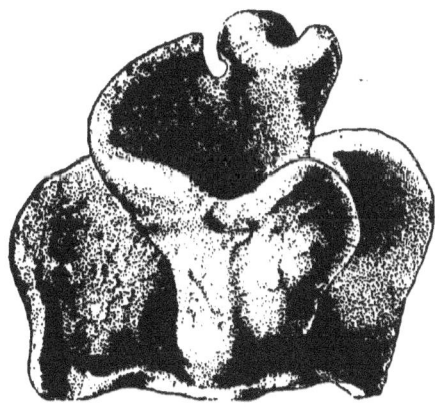

Fig. 533. — Vetulina stalactites. ¼.

ularly placed pores. The arrangement of the calcareous skeleton recalls to a certain extent that of the Hexactinellidæ. The genus Vetulina was previously known only from the Jura; it was quite commonly dredged off Barbados in 100 fathoms. An allied sponge, from 292 fathoms off Morro Light, Havana, is the little pear-shaped *Collinella inscripta* (Fig. 534), of which the fossil precursor may have been Trachysycon. Resembling *Sulcastrella clausa* (Fig. 535), dredged off Sand Key in 129

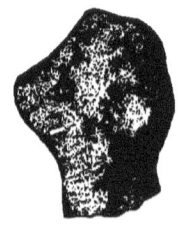

Fig. 534. — Collinella inscripta. ⅔.   Fig. 535. — Sulcastrella clausa. ⅔.

fathoms, are a number of cretaceous sponges to which Zittel has given the name of Astrobolia.

Cup-shaped sponges having bunches of spicules scattered over the surface, like Setidium (Fig. 536), are unusual among the

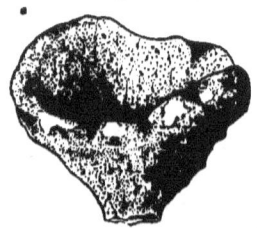

Fig. 536. — Setidium obtectum. ⅔.

Lithistidæ. An interesting lithistid is a small form, *Tremaulidium geminum* (Fig. 537), in which the pores are replaced by an inward tubular extension of the cuticle.

Fig. 537. — Tremaulidium geminum. ¼.

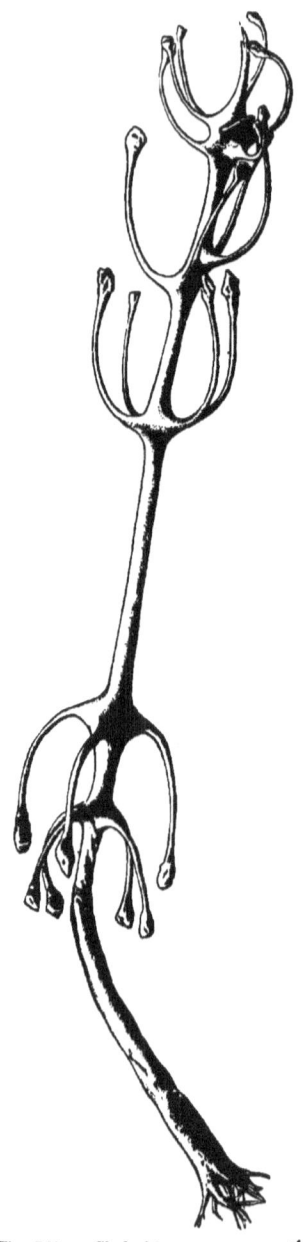

Fig. 541. — Cladorhiza concrescens. ½.

To the group of Tetractinellidæ belongs one of the most characteristic of the deep-sea sponges, *Tisiphonia fenestrata* (Fig. 538), of very variable appearance, with one or more afferent openings. These are specially protected in the allied *Fangophilina submersa* (Figs. 539, 539 a) by a tuft, which serves to fix it loosely in the mud. Closely allied to Lovén's

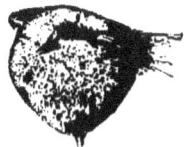

Fig. 538. — Tisiphonia fenestrata. ⅔.

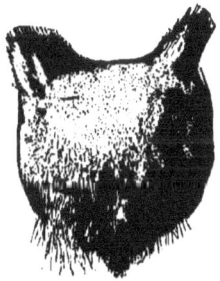

Fig. 539. Fangophilina submersa. ⅔. Fig. 539 a.

Hyalonema boreale is *Stylorhiza stipitata*. (Fig. 540.) Fragments and moderately complete specimens of Cladorhiza (Fig. 541) were not uncommon in the deeper dredgings of the "Blake." They are sponges with a long stem ending in ramifying roots deeply sunk in the mud. The stem has nodes with four to six club-shaped appendages. As Thomson has noticed, they evidently often cover, like bushes, extensive tracts of the bottom.

Among the Monactinellidæ we may mention Rhizochalina, which grows up between masses of coral and tubes of annelids, so as to be freely washed by water; also a very graceful branching form, *Phakellia tenax*.

Fig. 540. — Stylorhiza stipitata. ⅔.

178 THREE CRUISES OF THE "BLAKE."

(Fig. 542.) Nearly all the specimens of *Cribrella hospitalis*

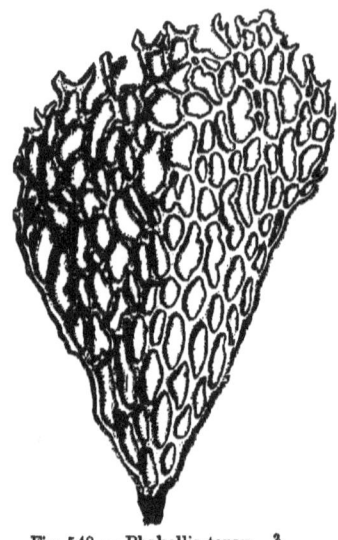

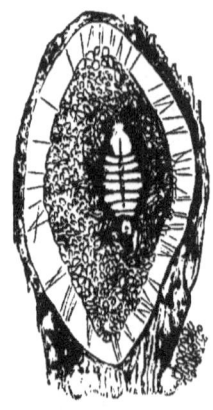

Fig. 543. — Blind Isopod parasitic of Cribrella hospitalis, magnified. (Schmidt.)

Fig. 542. — Phakellia tenax. ⅔.

are occupied by a parasitic blind isopod. (Fig. 543.) *Schmidtia aulopora* (Fig. 544) represents a widely distributed West Indian

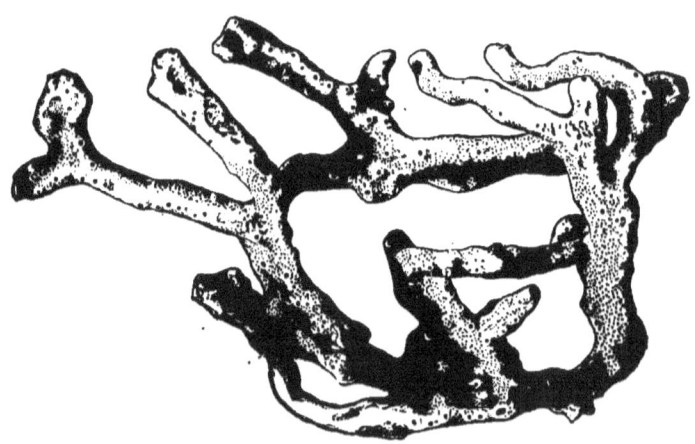

Fig. 544. — Schmidtia aulopora. ½.

form : a thick, coarse, smooth sheet, with stout branches, either round or angular.

One of the most abundant sponges is the small *Radiella sol* (Fig. 545), which extends to over 1,000 fathoms in depth. Its general appearance is that of a segment of a sphere surrounded by a fringe of needles, with a layer of larger needles radiating from the centre of the disk and forming the base.

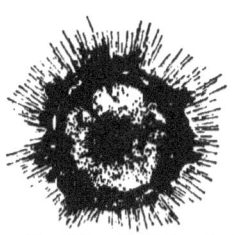

Fig. 545. — Radiella sol. ⅔.

LIST OF FIGURES.

INTRODUCTION.

FIGURE   VOL. I.   PAGE

A. Track of the "Blake" (U. S. C. S.[1]). In pocket at end of Vol. I.
B. Morro Light and Castle, Havana, with modern limestone terrace in foreground, from a photograph xii
C. Western Slope of St. Kitts, from a photograph xv
D. Pitons of St. Lucia, from a photograph xvii
E. Kingstown, St. Vincent, from a photograph xviii

I. EQUIPMENT OF THE "BLAKE." (Figs. 1–33.)

Frontispiece. U. S. Coast and Geodetic Survey Steamer "Blake" (U. S. C. S.).
1. Brooke's detacher (Sigsbee) 3
2. Sir William Thomson's sounding machine (Thomson) 4
3. Sigsbee's sounding machine (Sigsbee) 6, 7
4. Sigsbee's sounding machine, profile (Sigsbee) 7
5. Sigsbee's sounding machine, end view (Sigsbee) 8
6. Sigsbee's detacher and Belknap sounding cylinder (Sigsbee) . . . 10
7. Section of Fig. 6 (Sigsbee) 10
8. Belknap's cylinder detached from shot (Sigsbee) 10
9. Comparative size of hemp and wire used for sounding (Sigsbee) . . 12
10. Sounding lead with Stellwagen cup (Sigsbee) 12
11. Miller-Casella thermometer, ½ 15
12. Negretti-Zambra thermometer, Italian model 16
13. Siemens's sinker and resistance coil (Bartlett, U. S. C. S.) . . . 17
14. Wheatstone's Bridge (Bartlett, U. S. C. S.) 18
15. Siemens's deep-sea thermometer (Bartlett, U. S. C. S.) 19
16. Hilgard's salinometer (Sigsbee) 21
17. Sigsbee's water cup (Sigsbee) 23
18. Section of Sigsbee's water cup (Sigsbee) 23
19. Mode of attachment of cups and thermometer (Sigsbee) 23
20. Miller-Casella thermometer in protecting case, with Sigsbee's attachment (Sigsbee) 23
21. O. F. Müller's dredge (Thomson) 24
22. "Blake" dredge frame 24
23. "Blake" dredge 24
24. Bar and tangles 25
25. "Blake" trawl (Sigsbee) 26, 27
26. "Blake" trawl 26

[1] The authority from which the figure is derived is inserted in parentheses. U. S. C. S. = United States Coast Survey.

LIST OF FIGURES.

FIGURE	VOL. I. PAGE
27. Comparative size of steel and hemp dredging ropes (Sigsbee) . . .	28
28. Accumulator (Sigsbee)	31
29. Deck of "Blake," showing winding reel, and winding engine, facing bow, dredging boom, ready for dredging (Sigsbee) . .	30, 31
29 a. "Blake" deck looking aft, ready for dredging, reeling engine, drum, leading block (Sigsbee)	32, 33
30. Plan of deck of "Blake" (Sigsbee)	33
31. Scoop net	35
32. Tow net	35
33. Sigsbee's gravitating trap (Sigsbee)	36

III. THE FLORIDA REEFS. (Figs. 34–54.)

34. Straits of Florida, from Coast Survey charts	52
35. Distant mangrove islands seen from Mangrove Beach north of slaughter house, Key West	54
36. Coral breccia	54
37. Coral oölite	54
38. Map of the Tortugas, from Coast Survey charts	58
39. Barbados terraces	65
40. Sections across the Peninsula of Florida	64, 65
41. Sections across the Florida Reefs, as numbered on Fig. 34 . . .	66
42. Coral rock beach, north of Fort Taylor, exposed to the action of waves of channel between outer reef and Key West . . .	68, 69
43. Lagoon formed by promontories and low islands covered with mangroves, northeast shore of Key West Island	70, 71
44. Alacran Reef, Marquesas Keys, and sections across the reef and keys	72, 73
45. Pelagic Porites embryo, greatly magnified	74
46. Terraces near Fort Charles Light House, Barbados	79
47. Madrepora prolifera (Agassiz)	80, 81
48. Porites clavaria (Agassiz)	82
49. Udotea flabellata (Agassiz)	82, 83
50. Halimeda tridens (Agassiz)	82, 83
51. Madrepora palmata (Agassiz)	84, 85
52. Rhipidigorgia flabellum (Agassiz)	86, 87
53. Coral sand beach south of Navy Depot, Key West	88, 89
54. Orbicella annularis (Agassiz)	88

IV. TOPOGRAPHY OF THE EASTERN COAST OF THE NORTH AMERICAN CONTINENT. (Figs. 55–61.)

55. Atlantic basin, along Eastern United States, from model constructed by the U. S. Coast Survey and Hydrographic Office . . .	94
56. Contour map of the western part of the North Atlantic, U. S. Coast Survey and Hydrographic Office	94, 95
57. Contour map of the Caribbean Sea, prepared from data furnished by the U. S. Hydrographic Office, based on the deep-sea soundings of the U. S. C. S. S. "Blake" and the U. S. F. C. S. "Albatross"	98, 99
58. Contour map of the Windward Islands and the adjacent sea (U. S. C. S)	100
59. Contour map of the Gulf of Mexico (U. S. C. S.)	102

LIST OF FIGURES. 183

FIGURE		VOL. I. PAGE
60.	Contour map of the Gulf of Maine (U. S. C. S.)	103
61.	Contour chart of the bottom of the Atlantic, U. S. Hydrographic Office	108

VI. THE PERMANENCE OF CONTINENTS AND OF OCEANIC BASINS. (Fig. 62.)

	Geological map of the Eastern United States, after C. H. Hitchcock	129
62.	Archæan map of North America (Dana)	130

IX. THE PELAGIC FAUNA AND FLORA. (Figs. 63–139.)

63.	Arachnactis, embryo of Edwardsia, $\frac{1}{1}$	172
64.	Asteracanthion, starfish Pluteus, greatly magnified	172
64 a.	Strongylocentrotus, sea-urchin Pluteus, greatly magnified . . .	172
65.	Cyphonautes, embryo of Bryozoön, greatly magnified	172
66.	Littorina embryo, $\frac{1}{1}$	172
67.	Embryo of loligo, $\frac{2}{1}$	172
68.	Tornaria, embryo of Balanoglossus, greatly magnified	173
69.	Pilidium, larva of Nemertean, greatly magnified	173
70.	Leucodora larva, greatly magnified	173
71.	Polygordius larva, $\frac{2}{1}$	173
72.	Dactylopus larva, greatly magnified	173
73.	Barnacle, Balanus larva, greatly magnified	174
74.	Squilla embryo, $\frac{1}{1}$	174
75.	Hermit crab, Pagurus larva, greatly magnified	174
76.	Peneus embryo, greatly magnified	174
77.	Pelagic fish egg, $\frac{1}{1}$	175
78.	Plagusia. Pelagic symmetrical embryo of transparent flounder, $\frac{1}{1}$.	175
79.	Rhombodichthys, flounder stage of Plagusia, $\frac{1}{1}$	176
80.	Embryo fish, with lateral organs, $\frac{1}{1}$	176
81.	Phronima, $\frac{1}{1}$	176
82.	Leptocephalus, $\frac{2}{3}$	177
83.	Polyclonia frondosa, $\frac{1}{8}$ (Agassiz)	178
84.	Copepod, greatly magnified	178
85.	Mysis, $\frac{2}{1}$	179
86.	Porpita, $\frac{1}{1}$	180
87.	Physalia Arethusa, Portuguese man-of-war, $\frac{1}{2}$ (Agassiz) . . .	181
88.	Lepas anatifa, floating barnacle, slightly reduced	182
89.	Agalma elegans, $\frac{1}{2}$ (Fewkes)	182
90.	Cunina discoides, magnified	182
91.	Velella mutica, $\frac{1}{1}$	183
92.	Pterophysa grandis, $\frac{1}{12}$ (Fewkes)	184
93.	Linerges mercurius, $\frac{2}{3}$	186
94.	Glossocodon tenuirostris, magnified	186
95.	Fourth stage of Glossocodon tenuirostris larva, greatly magnified .	186
96.	Sixth stage of Glossocodon tenuirostris larva, greatly magnified .	186
97.	Janthina, slightly reduced	186
98.	Glaucus, enlarged	186
99.	Hyalea, $\frac{2}{1}$	187
100.	Atlanta, $\frac{1}{1}$	187
101.	Styliola, $\frac{2}{1}$	188
102.	Pleuropus, $\frac{1}{1}$	188

LIST OF FIGURES.

FIGURE		VOL. I. PAGE
103.	Tiedemannia, $\frac{1}{1}$	188
104.	Doliolum, $\frac{10}{1}$, from Newport	188
105.	Doliolum, $\frac{6}{1}$, from Florida	188
106.	Doliolum, $\frac{6}{1}$, from Florida, with well-developed zoöids	189
107.	Pyrosoma, $\frac{1}{2}$	189
108.	Appendicularia, greatly magnified	189
109.	House of Appendicularia, greatly magnified	189
110.	Solitary form of Salpa Caboti, $\frac{2}{1}\frac{5}{1}$	189
111.	Chain of Salpa Caboti, somewhat enlarged	190
112.	Firoloidea, $\frac{1}{2}$	191
113.	Argonauta, $\frac{2}{3}$ (Verrill)	191
114.	Spirula Peronii, $\frac{1}{1}$	191
115.	Shell of Argonauta, $\frac{2}{3}$ (Verrill)	191
116.	Sagitta, $\frac{1}{2}$	192
117.	Tomopteris, $\frac{2}{3}$	192
118.	Nebalia, $\frac{4}{1}$	193
119.	Globigerina bulloides, from the surface, $\frac{140}{1}$ (Challenger)	194
120.	Pterocanium charybdeum, highly magnified (Müller)	195
121.	Hialomma, $\frac{25}{1}$	195
122.	Sphærozoum, $\frac{12}{1}$	195
123.	Halobates wüllerstorffi, $\frac{4}{1}$ (Challenger)	179
124.	Hastigerina pelagica, $\frac{3}{1}\frac{5}{1}$ (Challenger)	196
125.	Noctiluca, magnified	197
126.	Eucope diaphana, $\frac{8}{1}$	197
127.	Lizzia grata, magnified	197
128.	Mnemiopsis Leidyi, somewhat reduced	199
129.	Zoea of Carcinus, greatly magnified	206
130.	Panopus embryo, greatly magnified	206
131.	Zoea of Porcellana, $\frac{4}{1}$	206
132.	Pelagic refuse, magnified	207
133.	Trichodesmium erythræum, $\frac{4}{1}$	209
134.	Coccosphere, $\frac{900}{1}$ (Challenger)	209
135.	Rhabdosphere, $\frac{140}{1}$ (Challenger)	209
136.	Gulf-weed, $\frac{1}{2}$	210
137.	Velella mutica, vascular canal filled with yellow cells (liver cells), magnified	214
137 a.	Single yellow cell, magnified	214
138.	Velella medusa, with yellow cells, magnified	215
139.	Phronima sedentaria, in its Doliolum house, $\frac{2}{3}$	215

X. TEMPERATURES OF THE CARIBBEAN, GULF OF MEXICO, AND WESTERN ATLANTIC. (Figs. 140-167.)

140.	Bottom temperatures, Western North Atlantic, J. R. Bartlett, U. S. N. (U. S. Hydrographic Office)	216, 217
141.	Temperature sections (curves)	217
142.	Bottom temperatures of the Atlantic (U. S. Hydrographic Office)	218

Figs. 143-167. TEMPERATURE SECTIONS (U. S. C. S.).

143.	Campeche Bank, Yucatan, to Cape San Antonio	219
144.	Virgin Gorda to Sombrero	221
145.	Point Rosalie, Dominica, to Point du Diable, Martinique	222

LIST OF FIGURES. 185

FIGURE		VOL. I. PAGE
146.	Salines Point, Martinique, to Point Hardy, St. Lucia	222
147.	Cape Moule à Chique, St. Lucia, to Tarratee Pt., St. Vincent	223
148.	Point Espada, San Domingo, to Point Jiguero, Porto Rico	223
149.	Guanos Point, Cuba, to Tortuga Island, San Domingo	225
150.	Morant Point, Jamaica, to Cape Tiburon, San Domingo	225
151.	Cape Cruz, Cuba, to Pedro Bank	226
152.	Santiago de Cuba to Morant Point, Jamaica	227
153.	Isle of Pines, Cuba, to Grand Cayman and Pedro Bank	228, 229
154.	Campeche Bank, Yucatan, to Loggerhead Key Light, Tortugas	230, 231
155.	Coast of Mexico, south of Vera Cruz, to Galveston Isl., Texas	228, 229
156.	Campeche Bank, Yucatan, to S. E. Pass Mississippi Delta	230, 231
157.	Garden Key Light, Tortugas, to Port Muriel, Cuba	231
158.	Cape Florida Light to Gun Key Light, Great Bahama Bank	232
159.	Jacksonville, Florida, to 350 miles east of Jacksonville	233
160.	North of Cape Cañaveral to 120 miles east of Cape Cañaveral	233
161.	South of Cape Cañaveral to 150 miles east of Cape Cañaveral	233
162.	Twenty to eighty miles from Cape Cañaveral	234
163.	Off St. Simon's Island, Georgia	235
164.	Off Santo Domingo to 275 miles in a N. E. direction	236
165.	Bahamas to Bermudas	237
166.	Forty miles north of Cape Hatteras to 110 miles east of same	237
167.	Off Montauk Point to the Bermudas	238

XI. THE GULF STREAM. (Figs. 168–176.)

168.	Surface temperature of the Atlantic in March (Krümmel)	241
169.	Surface temperature of the Atlantic in September (Krümmel)	243
170.	Temperature section Sambro Island, Halifax, to St. Thomas (Challenger)	244
171.	Ideal temperature section from the north to the south pole (J. J. Wild)	246
172.	Oceanic circulation, map of the 17th century, from Athanasii Kircheri, E. Soc. Jesu Mundus subterraneus editio tertia, Amstelodami, 1678	251
173.	Franklin's chart of the Gulf Stream (Kohl)	252
174.	Chart of the Gulf Stream showing its axis and limits from 1845–1860 (U. S. C. S.)	253
175.	Temperature section, Sandy Hook to the Bermudas (Challenger)	253
176.	Chart showing "Blake" Plateau and velocity of Gulf Stream (U. S. C. S., J. R. Bartlett, U. S. N.)	259

XII. SUBMARINE DEPOSITS. (Figs. 177–192.)

177.	Spherule of bronzite, $\frac{2}{1}$, from 3,500 fathoms, Central Pacific (Challenger)	262
178.	Black spherule, $\frac{3}{1}$ (Challenger)	263
179.	Spherule coated with magnetite, $\frac{2}{1}$ (Challenger)	263
180.	Pteropod ooze, $\frac{4}{1}$ (Murray and Renard)	264
181.	Globigerina ooze, $\frac{2}{1}$ (Murray and Renard)	265
182.	Globigerina slab, $\frac{1}{1}$, off Alligator Reef	265
183.	Radiolarian ooze, $\frac{7}{1}$ (Murray)	266
184.	Diatom ooze, $\frac{14}{1}$ (Murray)	266

FIGURE		VOL. I. PAGE
185.	Scalpellum Darwinii, attached to manganese nodule, $\frac{1}{1}$ (Challenger)	268
186.	Shark's tooth, Oxyrhina (Challenger)	268
187.	Ear-bone, Zyphius, 2,375 fathoms (Challenger)	268
188.	Section of manganese nodule showing tympanic bone of Mesoplodon, 2,600 fathoms (Challenger)	269
189.	Concretion, 333 fathoms, from Blake station 317, $\frac{2}{3}$	276
190.	Modern greensand, 142 fathoms, from Blake station 314, $\frac{15}{1}$	278
191.	Map showing distribution of bottom deposits	286
192.	Rock from Pourtalès Plateau	287

XIII. THE PHYSIOLOGY OF DEEP-SEA LIFE. (Figs. 193, 194.)

193.	Bunsen's apparatus, as modified by Jacobsen and Behrens (Vöringen exp.)	295
194.	Map of specific gravity of sea-water of the Western North Atlantic (Buchanan)	299

XV. SKETCHES OF THE CHARACTERISTIC DEEP-SEA TYPES. — FISHES. (Figs. 195–224.)

FIGURE		VOL. II. PAGE
195.	Sternoptyx diaphana, $\frac{4}{1}$	22
196.	Cyclothone lusca, $\frac{1}{1}$ (U. S. F. C.)	22
197.	Monolene atrimana, about $\frac{1}{1}$	24
198.	Barathronus bicolor, about $\frac{1}{5}$	25
199.	Barathrodemus manatinus, about $\frac{2}{3}$	25
200.	Aphyonus mollis, about $\frac{2}{5}$	25
201.	Ophidium cervinum, about $\frac{1}{2}$ (U. S. F. C.)	26
202.	Macrurus caribbæus, about $\frac{2}{3}$ (U. S. F. C.)	26, 27
203.	Bathygadus arcuatus, $\frac{1}{2}$ (U. S. F. C.)	26, 27
204.	Phycis Chesteri, $\frac{3}{5}$ (U. S. F. C.)	26, 27
205.	Bregmaceros atlanticus, $\frac{2}{1}$	27
206.	Plectromus suborbitalis, $\frac{1}{1}$ (U. S. F. C.)	28
207.	Callionymus Agassizii, about $\frac{1}{2}$	29
208.	Chiasmodon niger, about $\frac{1}{2}$ (U. S. F. C.)	29
209.	Peristedium longispatha, about $\frac{3}{5}$	30
210.	Nest of Pterophryne in Gulf-weed, about $\frac{1}{2}$	31
211.	Antennarius, $\frac{2}{3}$	31
212.	Alepocephalus Agassizii	32, 33
213.	Halosaurus macrochir, $\frac{1}{4}$	32, 33
214.	Chauliodes Sloani, $\frac{7}{12}$ (U. S. F. C.)	32, 33
215.	Ipnops Murrayi, about $\frac{1}{1}$	32
216.	Bathysaurus Agassizii, about $\frac{1}{4}$	32, 33
217.	Bathypterois quadrifilis, about $\frac{2}{3}$	32
218.	Benthosaurus grallator, $\frac{2}{5}$	32, 33
219.	Scopelus Mülleri, $\frac{1}{1}$ (U. S. F. C.)	33
220.	Malacosteus niger, $\frac{1}{1}$	35
221.	Synaphobranchus pinnatus, $\frac{1}{4}$ (U. S. F. C.)	34, 35
222.	Nemichthys scolopaceus, $\frac{1}{1}$ (U. S. F. C.)	34, 35
223.	Nettastoma procerum, $\frac{1}{1}$	34, 35
224.	Gastrostomus Bairdii, $\frac{1}{2}$ (U. S. F. C.)	34, 35

FIGURE — VOL. II. PAGE

XVI. CHARACTERISTIC DEEP-SEA TYPES. — CRUSTACEA. (Figs. 225–259.)

225. Anomalopus frontalis, $\frac{2.25}{1}$ (Alph. Milne-Edwards) 37
226. Anisonotus curvirostris, $\frac{2}{3}$ (A. Milne-Edwards) 38, 39
227. Pisolambrus nitidus, $\frac{2}{1}$ (A. Milne-Edwards) 38
228. Micropanope pugilator, $\frac{1.5}{1}$ (A. Milne-Edwards) 38
229. Acanthocarpus bispinosus, $\frac{1}{2}$ (A. Milne-Edwards) 38
230. Cyclodorippe nitida, $\frac{2}{1}$ (S. I. Smith) 39
231. Cyclodorippe nitida, front view, $\frac{4}{1}$ (S. I. Smith) 39
232. Lithodes Agassizii, $\frac{1}{3}$ (S. I. Smith) 40, 41
233. Xylopagurus rectus, $\frac{1}{1}$ (A. Milne-Edwards) 40
234. Xylopagurus rectus in its house, $\frac{1}{1}$ (A. Milne-Edwards) 40
235. Catapagurus Sharreri, with epizoanthus house, $\frac{2}{1}$ (S. I. Smith) . . 41
236. Catapagurus Sharreri, with house in base of actinia, $\frac{2}{1}$ (S. I. Smith) . 42
237. Munidopsis rostrata, $\frac{1}{1}$ (S. I. Smith) 42, 43
238. Munida, $\frac{1}{1}$ (S. I. Smith) 43
239. Pentacheles sculptus, $\frac{1}{1}$ (S. I. Smith) 42, 43
240. Nephropsis Agassizii, $\frac{2}{1}$ (S. I. Smith) 44
241. Phoberus cæcus, $\frac{1}{2}$ (A. Milne-Edwards) 44, 45
242. Glyphocrangon aculeatus, $\frac{1}{1}$ (S. I. Smith) 45
243. Sabinea princeps, $\frac{1}{1}$ (S. I. Smith) 45
244. Heterocarpus carinatus, $\frac{2}{1}$ (S. I. Smith) 46
245. Nematocarcinus ensiferus, $\frac{1.5}{1}$ (S. I. Smith) 46, 47
246. Acanthephyra Agassizii, $\frac{1}{1}$ (S. I. Smith) 46
247. Meningodora, $\frac{1}{1}$ (S. I. Smith) 47
248. Benthœcetes Bartletti, $\frac{1}{1}$ (S. I. Smith) 47
249. Gnathophausia Zœa, $\frac{2}{1}$ (A. Milne-Edwards) 48, 49
250. Syscenus infelix, about $\frac{1.5}{1}$ (Harger) 48
251. Rocinela oculata, $\frac{4}{1}$ (Harger) 48
252. Bathynomus giganteus, $\frac{1}{3}$ (A. Milne-Edwards) 48, 49
253. Epimeria loricata, $\frac{2}{1}$ (S. I. Smith) 49
254. Colossendeis colossea, $\frac{2}{3}$ (E. B. Wilson) 49
255. Scæorhynchus armatus, profile, legs omitted, $\frac{1}{1}$ (E. B. Wilson) . 50, 51
256. Scæorhynchus armatus, $\frac{1}{1}$ (E. B. Wilson) 50, 51
257. Scalpellum regium, $\frac{3}{4}$ (Hoek) 50
258. Verruca incerta, $\frac{8}{1}$ (Hoek) 50
259. Cypris, greatly magnified 51

XVII. CHARACTERISTIC DEEP-SEA TYPES. — WORMS. (Figs. 260–273.)

260. Hyalinœcia in its semi-transparent tube 52
261. Tubes of Diopatra Eschrichtii, $\frac{1}{1}$ (Ehlers[1]) 53
262. Tube of Diopatra glutinatrix 53
263. Tube of Hyalopomatus Langerhansi, $\frac{1}{1}$ 53
264. Maldane cuculligera, $\frac{2}{1}$ 54
265. Cirratulus melanacanthus, $\frac{2}{1}$ 54
266. Amphinome Pallasii, $\frac{3}{4}$ 54
267. Sthenelais simplex, $\frac{1.5}{1}$ 54
268. Anterior portion of Rhamphobrachium Agassizii, $\frac{2}{1}$ 55
269. Eunice conglomerans in its paper-like tube, $\frac{2}{3}$ 55
270. Diopatra glutinatrix, $\frac{2.5}{1}$ 56
271. Anterior part of Buskiella abyssorum, $\frac{1}{2}$ (McIntosh) 56

[1] The drawings of Annelids were all prepared under the supervision of Professor Ehlers.

LIST OF FIGURES.

FIGURE	VOL. II. PAGE
272. Pomalostegus stellatus, $\frac{2}{1}$	57
273. Hyalopomatus Langerhansi, $\frac{4}{1}$	57

XVIII. CHARACTERISTIC DEEP-SEA TYPES. — MOLLUSKS. (Figs. 274–337.)

Cephalopods.

274. Opisthoteuthis Agassizii, about $\frac{1}{2}$ (Verrill)	58
275. Nectoteuthis Pourtalesii, $\frac{4}{1}$ (Verrill)	59
276. Mastigoteuthis Agassizii, $\frac{1}{2}$ (Verrill)	59
277. Eledone verrucosa, $\frac{1}{2}$ (Verrill)	60
278. Alloposus mollis, $\frac{2}{3}$ (Verrill)	60
279. Benthoteuthis, $\frac{4}{1}$ (Verrill)	61
280. Spirula, $1\frac{1}{1}$ (Huxley)	61
281. Architeuthis princeps, $\frac{1}{18}$ (Verrill)	62, 63

Gasteropods and Lamellibranchs.

282. Pleurotoma (Ancistrosyrinx) elegans, $\frac{2}{1}$ (Dall [1])	66
283. Pleurotoma subgrundifera, about $\frac{2}{1}$	66
284. Dentalium perlongum, $\frac{4}{1}$	67
285. Calliostoma aurora, $\frac{2}{1}$	68
286. Gaza superba, $1\frac{1}{1}$	68
287. Leptothyra induta, $\frac{4}{1}$	69
288. Pleurotomaria Adansoniana, $\frac{2}{3}$	69
289. Pleurotomaria Quoyana, $\frac{1}{1}$	69
290. Marginella Watsoni, $\frac{2}{1}$	70
291. Ringicula leptocheila, $\frac{4}{1}$	70
292. Cancellaria Smithii, $\frac{3}{1}$	70
293. Mitra Swainsoni, $\frac{4}{1}$	70
294. Typhis longicornis, $\frac{2}{1}$	70
295. Triforis longissimus, $\frac{2}{1}$	71
296. Siliquaria modesta, $1\frac{1}{1}$	71
297. Vermetus erectus, $1\frac{1}{1}$	71
298. Pecten (Amusium) Dalli, $\frac{1}{1}$	72
299. Pecten phrygium, $\frac{4}{1}$	72
300. Cetoconcha bulla, $\frac{2}{1}$	72
301. Cetoconcha bulla, interior of valve, $\frac{2}{1}$	72
302. Cetochoncha elongata, $\frac{1}{1}$	72
303. Tindaria cytherea, $1\frac{1}{1}$	72
304. Cardium peramabilis, $\frac{4}{1}$	72
305. Modiola polita, $\frac{2}{1}$	73
306. Cuspidaria microrhina, $\frac{1}{1}$	73
307. Cuspidaria microrhina, valve seen from the hinge, $\frac{1}{1}$	74
308. Verticordia elegantissima, $\frac{1}{1}$	74
309. Verticordia perversa, $\frac{2}{1}$	74
310. Bushia elegans, $\frac{2}{1}$	74
311. Meiocardia Agassizii, $1\frac{1}{1}$	74
312. Vesicomya venusta, $1\frac{1}{1}$	75

Brachiopods.

313. Terebratula cubensis attached to piece of coral, $\frac{1}{1}$ (Davidson)	76
314. Terebratula cubensis, interior of valve, $1\frac{1}{1}$ (Davidson)	76
315. Waldheimia floridana, $\frac{1}{1}$	76

[1] Mr. Dall has supervised the drawings of the Gasteropods and Lamellibranchs.

LIST OF FIGURES.

FIGURE		VOL. II. PAGE
316.	Waldheimia floridana, interior of valve, $\frac{1}{1}$	76
317.	Terebratulina Cailleti, $\frac{2}{1}$ (Davidson)	77
318.	Terebratula caput-serpentis, one valve removed, showing arms, $\frac{1\cdot 2}{1}\frac{1}{1}$ (Davidson)	77
319.	Platydia anomioides, $\frac{4}{1}$ (Davidson)	77
320.	Platydia anomioides, one valve removed, showing arms, $\frac{5}{1}$ (Davidson)	77
321.	Crania Pourtalesii, $\frac{2\cdot 4}{1}$ (Dall)	77
322.	Discina atlantica, $\frac{2}{1}$ (Verrill)	77

Bryozoa.

323.	Crisia denticulata, $\frac{2}{1}$	78
323 a.	Crisia denticulata, magnified (Smitt)	78
324.	Diastopora repens, $\frac{1}{1}$	78
324 a.	Diastopora repens, magnified (Smitt)	78
325.	Farciminaria delicatissima, $\frac{1}{1}$ (Busk)	78
325 a.	Farciminaria delicatissima, magnified (Busk)	78
326.	Membranipora canariensis, $1\frac{1}{1}5$	79
327.	Cellularia cervicornis, $\frac{2}{1}$	79
327 a.	Cellularia cervicornis, magnified (Smitt)	79
328.	Caberea retiformis, magnified (Smitt)	80
329.	Vincularia abyssicola, $\frac{2}{1}$	80
330.	Escharipora stellata, $\frac{2}{1}$	80
330 a.	Escharipora stellata, magnified (Smitt)	80
331.	Tessadroma boreale, $\frac{3}{1}$	81
331 a.	Tessadroma boreale, magnified (Smitt)	81
332.	Hippothoa biaperta, $\frac{1}{1}$	81
332 a.	Hippothoa biaperta, greatly magnified (Smitt)	81
333.	Cellepora margaritacea, $\frac{2}{1}$	82
333 a.	Cellepora margaritacea, magnified (Smitt)	82
334.	Biflustra macrodon, $\frac{2}{1}$	82
335.	Porina subsulcata, $\frac{2}{1}$	82
336.	Retepora reticulata, $\frac{1}{1}$	82
337.	Heteropora, $\frac{4}{1}$	83

XIX. CHARACTERISTIC DEEP-SEA TYPES. — ECHINODERMS. (Figs. 338–421 a.)

Holothurians.

338.	Psolus tuberculosus, $\frac{2}{1}$ (Théel)	85
339.	Echinocucumis typica, $\frac{2}{1}$ (Théel)	85
340.	Stichopus natans, $\frac{1}{2}$ (Koren & Danielssen)	85
341.	Trochostoma arcticum, $\frac{2}{3}$ (Koren & Danielssen)	86
342.	Psychropotes longicauda, $\frac{1}{2}$ (Théel)	86
343.	Deima Blakei, $\frac{2}{3}$ (Théel)	86
344.	Benthodytes gigantea, $\frac{2}{5}$ (U. S. F. C.)	87
345.	Euphronides cornuta, $\frac{2}{3}$ (U. S. F. C.)	87
346.	Pælopatides confundens, $\frac{2}{3}$ (Théel)	88
347.	Ankyroderma affine, $\frac{2}{3}$ (Koren & Danielssen)	88

Sea-Urchins.

348.	Dorocidaris papillata, $\frac{2}{1}$	89
349.	Dorocidaris Blakei, $\frac{2}{1}$	90
350.	Porocidaris Sharreri, $\frac{2}{1}$	90, 91

Figure		Vol. II. Page
351.	Salenia Pattersoni, $\frac{2}{3}$	90, 91
352.	Salenia varispina, $\frac{1}{1}$	90
353.	Salenia Pattersoni, denuded test showing apical system, $\frac{3}{1}$	91
354.	Salenia varispina, young specimen, $\frac{6.0}{1}$	91
355.	Temnechinus maculatus, partly denuded of spines, $\frac{2.5}{1}$	92
356.	Trigonocidaris albida, partly denuded, $\frac{2}{1}$	92
357.	Podocidaris sculpta, partly denuded, $\frac{1.5}{1}$	92
358.	Cœlopleurus floridanus, $\frac{1}{2}$	93
359 *a*.	Asthenosoma hystrix, a few plates of test; $\frac{4}{1}$	94
359.	Asthenosoma hystrix, $\frac{3}{5}$	95
360.	Phormosoma placenta, $\frac{2}{3}$	95
361.	Aspidodiadema antillarum, $\frac{3}{1}$	96
362.	Aspidodiadema antillarum, magnified pedicellaria	96
363.	Hemipedina cubensis, partly denuded, $\frac{1}{1}$	97
364.	Neolampas rostellata, denuded, $\frac{3}{1}$	97
365.	Rhynchopygus caribæarum, denuded, $\frac{2.25}{1}$	97
366.	Neolampas rostellata, magnified apical system	98
367.	Macropneustes spatangoides, denuded, $\frac{3}{4}$	98
368.	Conolampas Sigsbei, $\frac{1}{1}$	99
369.	Hemiaster zonatus, $\frac{1}{1}$	100
370.	Hemiaster expergitus, denuded, $\frac{3}{1}$ (Lovén)	100
371.	Paleopneustes hystrix, $\frac{2}{3}$	100
372.	Palæotropus Josephinæ, denuded, profile, $\frac{2}{1}$	100
373.	Pourtalesia miranda, from below, $\frac{2}{1}$	101
374.	Pourtalesia miranda, profile, $\frac{2}{1}$	101
375.	Urechinus naresianus, from above, $\frac{1.25}{1}$	101
376.	Urechinus naresianus, profile, $\frac{1.25}{1}$	101

Starfishes.

Figure		Vol. II. Page
377.	Pentagonaster ternalis, $\frac{2}{3}$ (Perrier)	102
378.	Archaster pulcher, $\frac{1}{1}$ (Perrier)	103
379.	Anthenoides Peircei, $\frac{2}{3}$ (Perrier)	103
380.	Ctenaster spectabilis, disk and arm, $\frac{3}{4}$ (Perrier)	104
381.	Radiaster elegans, from below, $\frac{3}{4}$ (Perrier)	104
382.	Zoroaster Ackleyi, $\frac{3}{4}$ (Perrier)	105
383.	Hymenodiscus Agassizii, $\frac{1.5}{1}$ (Perrier)	106
384.	Hymenodiscus Agassizii, disk and base of arms, from below, $\frac{3}{1}$ (Perrier)	106
385.	Hymenodiscus Agassizii, magnified spine (Perrier)	106
386.	Archaster mirabilis, $\frac{1}{1}$ (Perrier)	107
387.	Brisinga coronata, disk, base of arms, and single arm, $\frac{2}{3}$ (Sars)	108

Ophiurans.[1]

Figure		Vol. II. Page
388.	Astrophyton cœcilia, $\frac{1}{2}$	110, 111
389.	Ophiocreas spinulosus, $\frac{1}{2}$	109
390.	Ophiozona nivea, $\frac{2}{1}$	110
391.	Ophiophyllum petilum, $\frac{4}{1}$	110
392.	Ophiocamax hystrix, $\frac{2}{1}$	110, 111
393.	Ophiopæpale Goësiana, $\frac{4}{1}$	111
394.	Ophiura Elaps, $\frac{1.25}{1}$	111
395.	Ophioconis miliaria, $\frac{6}{1}$	112

[1] Mr. Lyman has kindly supervised the drawings of Ophiurans.

LIST OF FIGURES.

Figure		Vol. II. Page
396.	Ophiomusium planum, $\frac{1}{1}$	112
397.	Ophiomyces frutectosus, $\frac{2}{1}$	113
398.	Ophiomastus secundus, $\frac{0}{1}$	113
399.	Sigsbeia murrhina, $\frac{1}{1}$	114
400.	Astrocnida isidis, $\frac{1}{1}$	115
401.	Ophiolipus Agassizii, $1, \frac{5}{1}$	115
402.	Ophiohelus umbella, part of disk and base of two arms, $\frac{8}{1}$ (Lyman)	116
403.	Ophiohelus umbella, bunch of umbrella-shaped spines, $\frac{60}{1}$ (Lyman)	116

Crinoids.

404.	Metacrinus angulatus, $\frac{1}{2}$ (Carpenter)	117
405.	Pentacrinus asterius, $\frac{1}{4}$ (Carpenter)	116, 117
406.	Pentacrinus stage of Actinometra meridionalis, magnified	117
407.	Pentacrinus decorus, $\frac{2}{3}$ (Carpenter)	118, 119
408.	Pentacrinus decorus, youngest specimen of Pentacrinus obtained by "Blake" $\frac{2}{1}$ (Carpenter)	118
409.	Pentacrinus Mülleri, $\frac{4}{5}$ (Carpenter)	119
410.	Pentacrinus Blakei, $\frac{2}{4}$ (Carpenter)	118, 119
411.	Rhizocrinus lofotensis, $\frac{2}{1}$ (Sars & Carpenter)	121
412.	Rhizocrinus Rawsoni, $\frac{1}{1}$ (Carpenter)	121
413.	Adult Holopus Rangi, $1, \frac{4}{1}$ (Carpenter)	123
414.	Half-grown Holopus Rangi, $\frac{6}{1}$ (Carpenter)	123
415.	Young Holopus Rangi, $1, \frac{0}{1}$	124
416.	Atelecrinus, $\frac{2}{1}$ (Carpenter)	124
417.	Antedon spinifera, $\frac{1}{1}$	125
418.	Actinometra pulchella, $\frac{1}{1}$	126
419.	Myzostoma filicauda, $1, \frac{0}{1}$ (Von Graff)	127
420.	Myzostoma Agassizii, $1, \frac{2}{1}$ (Von Graff)	127
421.	Cyst of Myzostoma cysticolum, $\frac{4}{1}$ (Von Graff)	127
421 a.	Cyst of Myzostoma cysticolum, parasite of Actinometra meridionalis, $\frac{4}{1}$ (Von Graff)	127

XX. CHARACTERISTIC DEEP-SEA TYPES. — ACALEPHS. (Figs. 422–448.)

Ctenophoræ and Hydromedusæ.

422.	Ptychogena lactea, $\frac{1}{1}$	128
423.	Ocyroë maculata, $\frac{1}{1}$	129
424.	Eucharis multicornis, $\frac{1}{2}$ (Chun)	130
425.	Dodecabostrycha dubia, $\frac{1}{2}$	131
426.	Periphylla hyacinthina, $\frac{2}{4}$ (Fewkes)	132
427.	Atolla Bairdii, $\frac{2}{4}$ (Fewkes)	133
428.	Agalma Okenii, $\frac{1}{4}$ (Fewkes)	134
429.	Gleba hippopus, $\frac{2}{4}$ (Fewkes)	134
430.	Diphyes acuminata, $\frac{2}{4}$ (Fewkes)	135
431.	Aglaophenia bispinosa, $\frac{2}{4}$ (Allman)	136, 137
432.	Aglaophenia bispinosa, magnified corbula (Allman)	136
433.	Aglaophenia bispinosa, lower part of stem of Fig. 431 (Allman)	136, 137
434.	Cryptolaria conferta, $\frac{1}{1}$ (Allman)	136
435.	Cryptolaria conferta, magnified, fusiform shaped bodies (Allman)	136
436.	Cladocarpus paradisea, $\frac{1}{1}$ (Allman)	136, 137
437.	Hippurella annulata, $\frac{2}{3}$ (Fewkes)	136, 137

FIGURE VOL. II. PAGE
438. Callicarpa gracilis, $\frac{1}{1}$ (Fewkes) 136, 137
439. Callicarpa gracilis, magnified corbula (Fewkes) 136, 137
440. Pleurocarpa ramosa, branch modified to corbula, magnified (Fewkes) . 137

Hydrocorallinæ.

441. Animal of Millepora alcicornis, $\frac{25}{1}$ (Agassiz) 138
442. Millepora nodosa, Dactylozoid Gastrozoid, magnified (Moseley) . . 138
443. Millepora alcicornis, $\frac{2}{3}$ (Agassiz) 138, 139
444. Pliobothrus symmetricus, $\frac{1}{1}$ (Pourtalès) 139
445. Cryptohelia Peircei, part of branch, $\frac{4}{1}$ (Pourtalès) 139
445 *a*. Cryptohelia Peircei, $\frac{1}{1}$ (Pourtalès) 139
446. Stylaster filogranus, $\frac{1}{1}$ (Pourtalès) 140
447. Distichopora foliacea, $\frac{1}{1}$ (Pourtalès) 140
448. Allopora miniacea, $\frac{1}{1}$ (Pourtalès) 141

XXI. CHARACTERISTIC DEEP-SEA TYPES. — POLYPS. (Figs. 449–483.)

Halcyonoids and Actinoids.

449. Umbellula Güntheri, $\frac{1}{1}$ 142, 143
450. Pennatula aculeata, $\frac{3}{4}$ (Koren & Danielssen) 142, 143
451. Kophobelemnon scabrum, $\frac{1}{1}$ (Verrill) 142
452. Anthoptilum Thomsoni, $\frac{1}{2}$ (Kölliker) 142, 143
453. Balticina finmarchica, $\frac{1}{4}$ (Koren & Danielssen) 142, 143
454. Actinauge nexilis, $\frac{3}{4}$ (Verrill) 143
455. Dasygorgia Agassizii, $\frac{1}{1}$ (Verrill) 143
456. Chrysogorgia, $\frac{1}{2}$ 144
456 *a*. Iridogorgia Pourtalesii, $\frac{1}{3}$ 145
457. Acanella Normani, $\frac{1}{2}$ 145, 146
458. Primnoa Pourtalesii, $\frac{3}{4}$ (Verrill) 146
459. Calyptrophora, $\frac{1}{2}$ 145, 146
460. Sagartia abyssicola, $\frac{3}{4}$ (Verrill) 147
461. Actinauge nodosa, $\frac{1}{3}$ (Verrill) 147
462. Caryophyllia communis, from above, $\frac{1}{1}$ (Pourtalès) 148
462 *a*. Caryophyllia communis, profile, $\frac{1}{1}$ (Pourtalès) 148
463. Stenocyathus vermiformis, $\frac{1}{1}$ (Pourtalès) 148
464. Thecocyathus cylindraceus, $\frac{3}{1}$ (Pourtalès) 149
465, 465 *a*, *c*. Deltocyathus italicus, three varieties, from above, $\frac{2}{1}$ (Pourtalès) 149
465 *b*. Deltocyathus italicus, from below, $\frac{2}{1}$ (Pourtalès) 149
465 *d*. Deltocyathus italicus, profile, $\frac{2}{1}$ (Pourtalès) 149
466. Paracyathus confertus, $\frac{1}{1}$ (Pourtalès) 150
467. Stephanotrochus diadema, $\frac{1}{1}$ (Pourtalès) 150
468. Flabellum Moseleyi, profile, $\frac{1}{1}$ (Pourtalès) 150
468 *a*. Flabellum Moseleyi, from above $\frac{1}{1}$ (Pourtalès) 150
469. Desmophyllum Riisei, $\frac{1}{1}$ (Pourtalès) 151
470. Desmophyllum solidum, $\frac{2}{1}$ (Pourtalès) 151
471. Rhizotrochus fragilis, $\frac{2}{1}$ (Pourtalès) 151
472. Lophohelia prolifera, $\frac{1}{1}$ (Pourtalès) 151
473. Amphihelia rostrata, $\frac{1}{1}$ 152
474. Axohelia mirabilis, $\frac{1}{2}$ (Pourtalès) 153
475. Thecopsammia socialis, $\frac{15}{1}$ (Pourtalès) 153

LIST OF FIGURES.

FIGURE		VOL. II. PAGE
476.	Fungia symmetrica, $\frac{2}{3}$ (Pourtalès)	153
477.	Diascris crispa, $\frac{1}{2}$ (Pourtalès)	153
478.	Antillia explanata, $\frac{3}{4}$ (Pourtalès)	154
479.	Leptoncmus discus, $\frac{6}{6}$ (Challenger)	154
480.	Haplophyllia paradoxa, $\frac{1}{4}$ (Pourtalès)	154
481.	Haplophyllia paradoxa, $\frac{2}{3}$ (Pourtalès)	155
482.	Antipathes spiralis, $\frac{1}{1}^a$ (Pourtalès)	155
483.	Antiphathes columnaris, $\frac{4}{5}$ (Pourtalès)	155

XXII. CHARACTERISTIC DEEP-SEA TYPES.— RHIZOPODS. (Figs. 484–519.)

484, 484 a.	Biloculina ringens, $\frac{6}{1}$ (Goës)	159
484 b.	Biloculina ringens, $\frac{1}{1}^a$ (Goës)	159
485.	Biloculina tenera, with expanded pseudopodia, $\frac{1}{1}^a$ (Schultze)	160
486.	Orbiculina adunca, $\frac{1}{1}^a$ (Brady)	160
487.	Orbiculina adunca, young, $\frac{2}{1}^a$ (Brady)	160
488.	Cornuspira foliacea, $\frac{6}{1}$ (Goës)	161
489.	Astrorhiza limicola, $\frac{4}{1}$ (Brady)	161
490.	Sorosphæra confusa, $\frac{1}{1}^a$ (Brady)	162
491, 491 a.	Hyperammina elongata, $\frac{7}{1}$ (Goës)	162
492.	Rhabdammina abyssorum, $\frac{6}{1}$ (Brady)	162
493.	Rhabdammina abyssorum, $\frac{1}{1}$ (Brady)	163
494.	Rhabdammina linearis, $\frac{6}{1}$ (Brady)	163
495, 495 a.	Reophax scorpiurus, $\frac{6}{1}$ (Goës)	163
496.	Thurammina papillata, $\frac{5}{1}^a$ (Brady)	164
497.	Ammodiscus tenuis, $\frac{1}{1}^a$ (Brady)	164
498.	Cyclammina cancellata, $\frac{3}{1}^a$ (Brady)	164
499.	Cyclammina cancellata, $\frac{2}{1}^a$ (Brady)	164
500.	Textularia sagittula, $\frac{1}{1}^a$ (Goës)	164
501, 501 a.	Textularia trochus, $\frac{6}{1}$ (Goës)	165
502, 502 a.	Valvulina triangularis, $\frac{6}{1}$ (Goës)	165
503.	Lagena distoma, $\frac{2}{1}^p$ (Brady)	165
504.	Nodosaria radicula, $\frac{2}{1}^a$ (Goës)	166
505.	Nodosaria communis, $\frac{4}{1}$ (Goës)	166
506.	Cristellaria crepidula, $\frac{5}{1}^a$ (Goës)	166
507.	Cristellaria calcar, $\frac{6}{1}$ (Goës)	166
508, 508 a.	Sagrina dimorpha, $\frac{6}{1}$ (Goës)	166
509.	Polymorphina ovata, $\frac{4}{1}^a$ (Brady)	166
510.	Orbulina universa, surface, $\frac{5}{1}^a$ (Challenger)	166, 167
511, 511 a, b.	Globigerina bulloides, various types of tests, $\frac{1}{1}^b$ (Goës)	167
512.	Orbulina universa, $\frac{1}{1}^p$ (Goës)	167
513.	Cymbalopora bulloides, $\frac{4}{1}^b$ (Challenger)	168
514.	Carpenteria balaniformis, $\frac{6}{1}$ (Goës)	168
515, 515 a.	Pulvinulina auricula, $\frac{1}{1}^b$ (Goës)	169
516.	Pulvinulina Menardii, $\frac{1}{1}^a$ (Goës)	169
517.	Pulvinulina Menardii, $\frac{1}{1}^a$ (Goës)	169
518.	Truncatulina Ungeriana, $\frac{1}{1}^a$ (Goës)	169
519.	Polytrema miniaceum, $\frac{1}{1}^a$	169

XXIII. CHARACTERISTIC DEEP-SEA TYPES.— SPONGES. (Figs. 520–545.)

520.	Farrea facunda, $\frac{2}{3}$	171
521.	Lefroyella decora, $\frac{2}{3}$	171

FIGURE		VOL. II. PAGE
522.	Aphrocallistes Bocagei, ^{3}	172
523.	Dactylocalyx pumiceus, $\frac{1}{2}$	172
524.	Regadella phœnix, $\frac{1}{4}$	173, 174
525.	Euplectella Jovis, $\frac{1}{2}$	173, 174
526.	Hyalonema Sieboldii, $\frac{2}{3}$	173
527.	Spicules of Japanese Hyalonema, with encrusting polyps, representing axis of Gorgonia, $\frac{1}{3}$	173
528.	Japanese Hyalonema, showing the siliceous cable and its parasitic polyps, $\frac{1}{2}$	174, 175
529.	Asconema setubalense, $\frac{1}{4}$ (Filhol. "Talisman" Ex.)	174, 175
530.	Pheronema Annæ, $\frac{2}{3}$	174, 175
531.	Holtenia Pourtalesii, $\frac{2}{3}$	174
531 *a*–531 *c*.	Spicules of Holtenia Pourtalesii (Schmidt)	174, 175
532.	Vetulina stalactites, spicules greatly magnified (Schmidt)	175
533.	Vetulina stalactites, $\frac{3}{4}$	175
534.	Collinella inscripta, $\frac{3}{4}$	176
535.	Sulcastrella clausa, $\frac{2}{3}$	176
536.	Setidium obtectum, $\frac{2}{3}$	176
537.	Tremaulidium geminum, $\frac{1}{1}$	176
538.	Tisiphonia fenestrata, $\frac{2}{3}$	177
539.	Fangophilina submersa, $\frac{2}{3}$	177
539 *a*.	Section through Fig. 539, $\frac{2}{3}$	177
540.	Stylorhiza stipitata, $\frac{2}{3}$	177
541.	Cladorhiza concrescens, $\frac{1}{2}$	176, 177
542.	Phakellia tenax, $\frac{2}{3}$	178
543.	Blind Isopod parasitic of Cribrella hospitalis, magnified (Schmidt)	178
544.	Schmidtia aulopora, $\frac{1}{2}$	178
545.	Radiella sol, $\frac{2}{3}$	179

INDEX.

AHYLA trigona, ii. 135.
Abyssal deposits, far from continents, i. 261.
   position of, i. 140.
Abyssal families, offshoots of free swimming, ii. 30.
Abyssal fauna, general character of, ii. 2.
   Lovén on derivation of, i. 155.
   Moseley on age of, i. 155.
   Moseley on derivation of, i. 156.
   Perrier on derivation of, i. 155.
Abyssal invertebrates and fishes, huge eyes of, i. 307.
Abyssal mollusks, enemies of, ii. 65.
   general character of, ii. 62.
   limited by cold, ii. 64.
   limited variety of forms of, ii. 64.
   points of attachment of, ii. 64.
Abyssal realm, character of, ii. 1.
Abyssal shells, colors of, ii. 63.
   delicacy of, ii 63
   ornamentation of, ii. 63.
Acalephs, ii. 128.
   occurrence of on the surface, i. 177.
   pelagic, i. 185.
   phosphorescence of, i. 197.
   report on by J. W. Fewkes, i. xxi.
   shoals of, i. 186.
   swimming near the bottom, i. 202.
Acanella Normani, ii. 145.
Acanthephyra Agassizii, ii. 46.
Acanthocarpus Alexandri, ii. 38.
Acanthocarpus bispinosus, ii. 38.
Acanthodromia, ii. 40.
Acanthometridæ, i. 195.
Accumulator, i. 30.
Ackley, S. M., i. viii, 32.
Actinauge, ii. 143.
Actinauge nodosa, ii. 147.
Actiniæ, Hertwig on yellow cells of, i. 214.
   deep-sea attached to sea wands, ii. 143.
   incrusting masses of, ii. 148.
   phosphorescence of, ii. 147.

Actinoids, ii. 142.
   color of deep-sea, i. 312.
Actinometra meridionalis, ii. 117, 127.
Actinometra, pentacrinus stage of, ii. 117.
Actinometra pulchella, ii. 125.
Adamsia associated with Catapagurus, ii. 41.
Adrians, " Ingegerd " and " Gladan " Expedition, i. 40.
Ægean Sea, Edward Forbes on, i. 40.
Agalma, i. 121, 181.
   at Newport, ii. 133.
   Okenii, ii. 133.
Agaricia, i. 55.
Agassiz, A. Report on Coral Reefs, i. xxi.
   Report on Gulf Stream, i. xxi.
   Report on Sea-Urchins, i. xxi.
   Report on Surface Fauna, i. xxi.
Agassiz, L., i. vii, 285.
   on Coral Reef of Florida, i. 56.
   on movements of Rhizocrinus, ii. 122.
Agassizia, i. 159.
Aglaophenia bispinosa, ii. 136, 137.
Aglaophenia crenata, ii. 135.
Agonidæ, ii. 30.
Agonti, i. 114.
Alacran atoll, i. 70.
   not due to subsidence, i. 72.
Alacran, structure of reef, i. 71.
Alaminos, expedition of, i. 251.
" Albatross," i. 50.
" Albatross," on Ridge between Santa Cruz and Porto Rico, i. 98, 112.
Alepocephalus Agassizii, ii. 32, 33.
Algæ, Berthold on bathymetrical range of, i. 312.
   calcareous, i. 312.
   parasitic, Cienkowsky on, i. 214.
Allman, G. J., on deep-sea hydroids, i. xxi; ii. 7.
Allopora miniacea, ii. 140.
Alloposus mollis, ii. 60.
Altitudes near shore lines, i. 132.
Amblyrhynchus, i. 115.

American continent from Huronian to Tertiary, i. 106.
Ammodiscus tenuis, ii. 164.
Ampharetidæ, tubes of, ii. 56.
Amphihelia rostrata, ii. 152.
Amphinome Pallasii, ii. 54.
Amphipods, ii. 40.
　Smith on tubes of, ii. 53.
Anadyomene, i. 82.
Ananchytidæ, i. 159; ii. 100.
Ancient types and oceanic dredgings, i. 155.
Anguilla, fossil mammals of caves of, i. 114.
Anisonotus curvirostris, ii. 37, 38, 39.
Ankyroderma affinis, ii. 88.
Annelids, bathymetrical range of, ii. 53.
　E. Ehlers report on, i. xxi.
　littoral groups of, ii. 53.
　parasitic in corals, ii. 156.
　parasitic on Antipathes, ii. 156.
Annelid tubes, composition of, ii. 53.
　covering large tracts, ii. 53.
　transported, ii. 54.
Anomalopus frontalis, ii. 37.
Anomura, ii. 39.
Antarctic regions, Alph. Milne-Edwards on fauna of, i. 121.
Antedon Hagenii, ii. 124.
Antedon Sarsii, ii. 118.
Antedon spinifera, ii. 125.
Antedonin, i. 309.
Antennarius, ii. 31.
Anthenoides Peircei, ii. 103.
Anthoptilum Thomsoni, ii. 142, 143.
Antigua, island of, i. xix.
Antillia explanata, ii. 154.
Antillean continent, i. 116.
Antipathes columnaris, ii. 155.
Antipathes spiralis, ii. 155.
Antipathidæ, ii. 155.
Antique types, i. 156.
Aphrocallistes Bocagei, ii. 172.
Aphyonus mollis, ii. 25.
Apiocrinidæ, ii. 116.
Apoda, ii. 84.
Appendicularia, i. 187.
Arago, i. 249.
Arbacia, i. 159.
Arbaciadæ, ii. 92.
　spines of, ii. 92.
Archaster mirabilis, ii. 102, 107.
Archaster pulcher, ii. 103.
Archasteridæ, ii. 102.
Archeocidaris, plates of, ii. 96.
Architeuthis princeps, ii. 62.
Arctic current, Bartlett on southern extension of, i. 270.

Arctic current, course of, i. 121.
　influence of, i. 134.
Arctic regions, warm climate of. i. 134.
Arctic species, cropping out of, i. 302.
Areas of depression, i. 126.
Areas of elevation, i. 126.
Argonauta, i. 191.
　Giglioli on, i. 193.
Argyope. i. 193.
Argyropelecus, ii. 22.
Ascidians, ii. 77.
Asiatic continent, relation of to East Indian Archipelago, i. 125.
Askonema setubalense, ii. 174, 175.
Aspidodiadema antillarum, ii. 96.
Aspidodiadema, sheathed pedicellariæ of, ii. 96.
Aspidosiphon, ii. 52.
Astacidea, ii. 43.
Astacus zaleucus, i. 308.
Astarte, ii. 73.
Asthenosoma, Grube on, ii. 94.
Asthenosoma hystrix, ii. 94, 95.
Asthenosoma, plates of test of, ii. 94.
　shape of, ii. 94.
　sheathed spines of, ii. 95.
Astrocnida, ii. 115.
Astrocnida isidis, ii. 5.
Astronyx Loveni, ii. 5.
Astropecten, ii. 103.
Astrophytidæ, ii. 109.
Astrophyton, ii. 114.
　from Baffin's Bay, i. 41.
　swimming of, i. 44.
Astrophyton cœcilia, ii. 110, 111.
Astrorhiza limicola, ii. 161.
Astrorhiza, Sundahl, Dr., on, ii. 161.
Astroschema, ii. 5.
Atelecrinus, ii. 124.
　a larval form of Comatula, ii. 126.
Athorybia formosa, ii. 133.
Atlanta, i. 187, 265.
Atlantic and Pacific continents, i. 123.
Atlantic and Pacific, former connection of, i. 92.
Atlantic, eastern basin of, i. 242.
　hydrographic character of, i. 242.
Atlantic ooze, Bailey on, i. 45.
　composition of, i. 150.
Atlantic slope, profile of, i. 134.
Atlantic, western basin of, i. 242.
Atlantis, i. 126.
Atolla Bairdii, ii. 132, 133.
Atolla Wyvillei, ii. 132.
Aurelia, i. 186.
Austins, the, on Pentacrinidæ, ii. 117.

INDEX. 197

Axohelia mirabilis, ii. 152, 153.
Ayres, i. 43.

Bache, A. D., i. 45.
   on Gulf Stream in 1845, i. 252.
Baer, K. E. v., on eastern extension of Gulf Stream, i. 252.
Bahama Bank, eastern slope of, i. 96, 104.
   formation of, i. 69.
   land shells of, i. 116.
   slope of, i. 288.
Bahama Banks, structure of, i. 75.
Bahama plateau, i. 75.
Bahia Honda, i. viii.
Bailey, i. 3.
   on Atlantic globigerinæ, i. 146.
   on Atlantic ooze, i. 45.
   on greensand of Zeuglodon limestone, i. 278.
Baird, S. F., U. S. Fish Commissioner, i. 50.
Balticina finmarchica, ii. 142, 143.
Barathrodemus manatinus, ii. 25.
Barathronus bicolor, ii. 25.
Barbados, island of, i. xix.
   limestone terraces of, i. 63.
Barbuda, island of, i. xix.
Bartlett Deep, i. 100, 226.
Bartlett, John R., i. viii.
   exploration of Caribbean, i 50.
   on current passing over ridge of windward passage, i. 202.
   on deep-sea sounding and dredging, i. 51.
   on path of warm water in Gulf of Mexico and Caribbean, i. 255.
   on Pentacrinus off Santiago de Cuba, ii. 6.
   on rip off Charleston, i. 254.
   on Sargassum, i. 211.
   on temperature sections of Caribbean and of Gulf Stream, i. 217.
   on temperature sections between the West India Islands, i. 218.
Barrett and Andrews, i. 43.
Barrett, survey of Alacran, i. 70.
Bathometer, Siemens, C. W., i. 6.
Bathybius, Haeckel on, i. 204.
Bathydoris abyssorum, ii. 62.
Bathygadus arcuatus, ii. 26, 27.
Bathymetrical faunal subdivisions, i. 162, 163.
Bathymetrical range, of corals, ii. 11.
   of crinoids, ii. 10.
   of crustacea, ii. 9.
   of fishes, ii. 8.
   of gorgonians, ii. 11.

Bathymetrical range, of mollusca, ii. 10.
   of ophiurans, i. 108.
   of sponges, ii. 11.
   of sea-urchins, ii. 10.
   of starfishes, ii. 10.
Bathyonomus giganteus, ii. 48, 40.
Bathypterois quadrifilis, ii. 32.
Bathysaurus Agassizii, ii. 32, 33.
Baur on range of pelagic animals, i. 200.
Behring Strait current, i. 243.
Belknap, Geo. E., i. 6.
   sounding cylinder, i. 3.
Bellerophon, i. 100.
Bemini, Straits of, i. 137.
   current flowing north through the, i. 234.
Benthœcetes Bartletti, ii. 47.
Benthodytes gigantea, ii. 87.
Benthosaurus grallator, ii. 32, 33.
Benthoteuthis, ii. 61.
Bergh, R., on Bathydoris abyssorum, ii. 62.
Bermuda sea-serpent, ii. 28.
Bermudas, not inhabited by Caribs, i. 118.
   origin of fauna and flora of, i. 119.
   recent origin of, i. 117.
   Rein on the, i. 80.
   to West India islands, ocean bed from, i. 93.
   vegetation of, i. 117.
Berthold, on bathymetrical range of algæ, i. 312.
Beryx splendens, ii. 27.
"Bibb," i. 51.
Biflustra macrodon, ii. 82.
Biloculina clay, ii. 160.
   Pourtalès on, ii. 160.
   Sars on, ii. 160.
Biloculina ringens, ii. 159.
Biloculina tenera, ii. 160.
Bird fauna of West Indies, i. 114.
Birds, effects of currents on distribution of, i. 120.
"Blake," i. 51.
"Blake," plan of deck of, i. 33.
   cruises of, i. viii, xi, xix, 38.
   rich dredgings by, off the West India islands, ii. 13.
   small size of, i. 32.
"Blake" dredges, i. 24.
Blake Plateau, i. 96, 135.
   deposits on steep slope of, i. 277.
   swept clean, i. 259.
Bland, on land shells of West Indies, i. 115.
Blennies, ii. 20.
Blind fishes, i. 308.
Blind invertebrates, i. 307.
Block Island soundings, i. 272.

198 INDEX.

Blue colors, absence of, i. 310.
Boguslawski on solids in ocean water, i. 129.
Bombay duck, ii. 34.
Boring mollusks and annelids, i. 55.
Bottom deposits, amount of carbonate of lime in, i. 275.
   change in character of, i. 277.
   color of, i. 280.
   how obtained, i. 262.
   land débris in, i. 291.
   north and south of Cape Hatteras, i. 270.
   of Antilles, i. 290.
   of Caribbean, i. 260, 288.
   of East Coast, Bailey sketch of, i. 269.
   of East Coast, Pourtalès sketch of, i. 269.
   of Gulf of Mexico, i. 260, 280.
   of Gulf Stream, i. 277.
   transition of, i. 267.
   vegetable matter in, i. 291.
Bottom ooze, cold, i. 303.
Bottom specimens, decrease of tints of, i. 281.
   examination of, i. 3.
Bottom temperatures of Caribbean, i. 218.
   of Caribbean and Gulf of Mexico, i. 245.
   of Gulf of Mexico, i. 218.
Bottom water, temperature of, i. 303.
Bourgueticrinus, i. 285.
Bourgueticrinus Hotessieri, Pourtalès on, ii. 120.
Boyle and Hooke, i. 39.
Brachiopods, ii. 75.
   character of, ii. 75.
   in palæozoic times, ii. 75.
   number of fossil species of, ii. 75.
   rare in collections, ii. 75.
   small number of recent species, ii. 75.
Brady on "Challenger" foraminifera, ii. 157.
Branching stars, ii. 109.
Brandt, on parasitic algæ, i. 214.
Brazilian current, i. 251.
Breccia beach, i. 87.
Bregmaceros atlanticus, ii. 27.
Bridgetown, i. xix.
Brisinga coronata, ii. 108, 109.
Brooke, John M., sounding apparatus and detacher, i. 3.
Brotulids, ii. 25.
Brown, Robert, i. 180.
Brownson, W. H., i. 93.
   deep-sea sounding by, i. 50, 238.
   on temperature sections in Western Atlantic, i. 221.
Bryozoa, ii. 78.

Bryozoa, association of with other animals, i. 215.
   forests of, i. 141.
   report on by Smitt, i. xxi.
Buchanan, J. Y., chemistry of sea water, i. 23.
   on specific gravity of ocean water, i. 208.
   on specific gravity of sea water of American coast, i. 209.
"Bulldog" machine, i. 42.
Bushia elegans, ii. 74.
Busk on Farciminaria, ii. 79.
Buskiella abyssorum, ii. 56.
Burrowing animals, diversity of colors in, i. 310.

Caberea retiformis, ii. 80.
Cadulus, ii. 67.
Calanus, i. 179, 193.
Callicarpa gracilis, ii. 137, 138.
Callionymus Agassizii, ii. 29.
Calliostoma aurora, ii. 68.
Calliostoma Bairdii, ii. 68.
Calliostoma psyche, ii. 68.
Calycophoræ, ii. 135.
Calyptrophora, ii. 146.
Cancellaria Smithii, ii. 70.
Cancroidea, ii. 37.
Cañon between Santa Cruz and St. Thomas, i. 98.
Carbonate of lime in bottom deposits, i. 271.
Carbonate of lime shells, solution of, i. 266.
Carbonic acid in sea water, determination of, i. 205.
Dittmar on, i. 297.
Cardium peramabilis, ii. 72, 73.
Caribbean, bottom deposits of, i. 280.
   calcareous ooze of, i. 290.
   density of, i. 300.
   eastern basin of, i. 98.
   gulf of the Pacific, i. 112.
   heaping up of water in, i. 248.
   superheated, i. 243.
   western basin of, i. 98.
Caribbean and Atlantic, connection of, i. 112.
Caribbean islands, affinity of fauna and flora, i. 114.
Carinaria, i. 191.
Carpenter, P. H., on stalked crinoids, ii. 10.
   on "Blake" crinoids, ii. 116.
   on "Challenger" Comatulæ, ii. 125.
Carpenter, W. B., i. 40, 285.
   on corallines, i. 166.
   on organic broth, i. 313.
   thermic theory of, i. 240.

Carpenter, W. B., and Thomson on phosphorescent animals, i. 308.
Carpenteria balaniformis, ii. 168.
Caryophyllia communis, ii. 148.
Castries, i. xvii.
Catapagurus Sharreri, ii. 41, 42.
Caudina, ii. 85.
Cellepora margaritacea, ii. 82.
Cellularia cervicornis, ii. 80.
Cellularians, allied to Australian types, ii. 80.
Central America in tertiary period, i. 116.
Centroceras, Pourtalès on, i. 313.
Centrophorus, ii. 30.
Centroscyllium Fabricii, ii. 30.
Cephalopods, ii. 58.
  at great depths, ii. 62.
  effect of pressure on, i. 304.
Ceratiidae, ii. 32.
Ceratodus, ii. 36.
Ceratoisis ornata, ii. 145.
Certes, on microbes at great depth, i. 314.
Cetoconcha bulla, ii. 72.
Cetoconcha elongata, ii. 72.
Chaetopods, deep-sea species of, ii. 55.
Chalk, chemical constitution of the, i. 146.
  composition of, i. 147.
  deep-sea animals in the, i. 146.
  forming at great depths near continents, i. 148.
  from New Zealand, i. 148.
  near reefs, i. 286.
  of New Britain, i. 148.
  off Nuevitas, i. 148.
Chalk marl, character of, i. 148.
"Challenger," i. 4.
  on cold bands of Agulhas current, i. 254.
  rich dredgings by, off Japan, ii. 13.
  saltest water found by, i. 209.
Challenger Deep, i. 106.
"Challenger" Expedition, i. 43.
Challenger ridge, i. 123, 164, 242.
Channel, between Cuba and Jamaica, i. 111.
  between Jamaica and San Domingo, i. 98.
  between San Domingo and Porto Rico, i. 98.
  between Santa Cruz and St. Thomas, i. xiv, 112.
Channels between the Virgin Islands, i. 98.
Chamisso, i. 180.
Charleston, S. C., i. xix.
Chauliodes Sloani, ii. 32, 33.
Chauliontidae, ii. 32.
Chaunax pictus, ii. 32.
Cheilostomata, ii. 70.
Chemical denudation, absence of, i. 104.
  M. Reade on, i. 128.

Chemical results, of "Challenger" Expedition, i. 294.
  of "Vöringen" Expedition, i. 294.
Chiasmodon niger, ii. 29.
Chilian plain, i. 120.
Chitonidae, ii. 67.
Chlamydoselachus, the frilled shark, ii. 36.
Chrysogorgidae, ii. 144.
Ciduridae, ii. 88.
Cidaris, i. 158.
Cienkowsky on parasitic algae, i. 214.
Cirratulus melanacanthus, ii. 54.
Cirripeds, abyssal, ii. 50.
Cladocarpus paradisea, ii. 137, 138.
Cladorhiza concrescens, ii. 176, 177.
Cladorhiza, Thomson on, ii. 177.
Claparède, i. 200.
Clarke, S. F., Report on Hydroids, i. xxi.
Clay bottom near Block Island, i. 272.
Clay lumps and concretions, i. 273.
Clement, C., i. 277.
Cleve on fossiliferous rocks of West Indies, i. 100.
Clio, i. 121.
Clypeastroids, tertiary, i. 150.
  absence of, in deep water, ii. 97.
Coast line, break in, i. 95.
Coast Survey office, i. xxi.
Coccoliths, i. 209.
Coccospheres, i. 209.
Coccospheres and rhabdospheres, C. Wyville Thomson on, i. 209.
Coelopleurus, i. 160.
Coelopleurus floridanus, ii. 93.
Collecting cylinder, specimens brought up by, i. 262.
Collections of "Blake," disposition of, i. xx.
Collinella inscripta, ii. 176.
Collozoum, i. 195.
Color, blue not a protective one, i. 310.
  of deep-sea crustacea, i. 312.
  of marine animals, i. 311.
Colors, Secchi on penetration of, i. 305.
Colossendeis colossea, ii. 49, 50.
Colossendeis macerrima, ii. 50.
Columbus, his theory of currents, i. 251.
  in northeast trades, i. 250.
  on Sargasso Sea, i. 213.
Comatulae, abundance of, ii. 118.
  bathymetrical range of, ii. 11.
  character of Caribbean, ii. 124.
Concretions, calcareous, off Barbados, i. 290.
  Clement, C., analysis of, i. 277.
Connection, between Caribbean and Atlantic, i. 112.
  between Caribbean and Pacific, i. 112.

Conoclypus Sigsbei, ii. 99.
Conolampas, ii. 97.
Conolampas Sigsbei, ii. 99.
Continental areas, deposits on, i. 140.
Continental belt, temperature of, i. 302.
Continental connections, extent of, i. 121.
Continental denudation, i. 282.
Continental fauna, i. 162.
  decrease of, i. 107.
Continental formations, i. 143.
Continental lands, height of, i. 126.
Continental line, ancient extension of, i. 136.
Continental masses, i. 126.
  effect on distribution of temperature, i. 248.
  Krümmel on elevation of, i. 126.
  nucleus of, i. 126.
Continental plateau, edge of, i. 108.
Continental shelf, i. 96.
  absence of argillaceous matter on, i. 274.
  sandy plain of, i. 272.
Continental slopes, abundance of animal life on, i. 107.
  fauna adjacent to, i. 106.
Continents and oceanic basins, Agassiz, L., on age of, i. 127.
  Carpenter, W. B., on age of, i. 127.
  Dana on age of, i. 127.
  Geikie, A., on great age of, i. 127.
  Guyot on great age of, i. 127.
  Thomson, Wyville, on age of, i. 127.
  Wallace on age of, i. 127.
Continents, permanence of, i. 125.
Convection through ocean water, i. 304.
Copepods, fertility of, i. 204.
  pelagic, i. 178.
Coquimbo, elevated coast near, i. 129.
Coquina of St. Augustine, i. 67, 68.
Coral bottom, extent of, i. 286.
Coral boulders, i. 55.
Coral breccia, cementation of, i. 54.
Corallines, i. 55.
Coral reef of Florida, Agassiz's theory of, i. 55.
  Leconte, J., on theory of, i. 55.
  report on by A. Agassiz, i. xxi.
  theory of E. B. Hunt, i. 55.
Coral reefs, ancient, near Havana, i. 71.
  Darwin's theory of, i. 55, 80.
  distribution of, i. 76, 286.
  effect of light on, i. 300.
  grinding and rehandling of material on, i. 55.
  of former geological periods, i. 109.

Coral reefs, northern extension of, i. 101.
  Semper on, i. 76.
  Studer on, i. 76.
Coral rock shores, undermining of, i. 87.
Corals, ii. 148.
  absence of simple species in Caribbean area, ii. 19.
  composition of, i. 148.
  depth of, affected by local causes, i. 74.
  known previous to "Blake" Expedition, ii. 7.
  limit to which they extend, i. 74.
  living on edge of Bahama Bank, i. 75.
  of Chagos Archipelago, i. 74.
  of miocene beds, ii. 19.
  Sharples, S. P., analysis of, i. 62.
Coral sand beach, i. 86.
Coral sand, held in suspension, i. 84.
  Wright, on slope of, i. 83.
Coral silt, carried along the bottom, i 84.
Corbulæ of Plumularidæ, ii. 137.
Corniferous bone beds, i. 145.
Cornuspira foliacea, ii. 161.
Corycodus bullatus, ii. 39.
Coryphæna, i. 193.
Coryphænoides, ii. 26.
Cosmopolitan species, i. 162.
Crangonidæ, ii. 45.
Crania Pourtalesii, ii. 77.
Cretaceous deposits of Isthmus of Panama, i. 113.
Cretaceous sea, Jeffreys on depth of, i. 146.
Cretaceous types, i. 151.
  in West Indian miocene, ii. 19.
Cribrella hospitalis, ii. 178.
Crinoid collection, disposition of, ii. 6.
Crinoids, ii. 116.
  color of deep-sea, i. 312.
  known previous to "Blake" Expedition, ii. 6.
  P. H. Carpenter on, i. xxi.
Crinoids and trilobites, great development of in Silurian, i. 155.
Crisia denticulata, ii. 78, 79.
Crisia eburnea, ii. 79.
Cristellaria calcar, ii. 160.
Cristellaria crepidula, ii. 160.
Croll, i. 247.
Cruises of the "Blake" in 1877, i. 50, 80.
Crustacea, ii. 37.
  bathymetrical range of, i. 169.
  habits of deep-sea, i. 311.
  knowledge of previous to "Blake" Expedition, ii. 4.
  living under moist stones, i. 153.
  modifications of, ii. 44.

Crustacea, organs of sight of, ii. 44.
　Alph. Milne-Edwards, report on, i. xxi.
　report on, by S. I. Smith, i. xxi.
Cryptohelia Peircei, ii. 139.
Cryptolaria conferta, ii. 136.
Ctenaster spectabilis, ii. 104.
Ctenophores, ii. 128.
　phosphorescence of, i. 174.
Ctenostomata, ii. 79.
Cuba, barrier reef of, i. 110.
　bottom on north shore of, i. 288.
　fringing reef of, i. 110.
Cunina, i. 182.
Currents, effect of in distribution of fauna, i. 92.
Currents and tides, effect of on topography, i. 104.
Currents of early geological periods, i. 154.
Cuspidaria microrhina, ii. 73, 74.
Cutlass fishes, ii. 28.
Cyanea, i. 186.
Cyclammina cancellata, ii. 164.
Cyclodorippe nitida, ii. 38.
Cyclopteridæ, ii. 28.
Cyclothone lusca, ii. 9, 22.
Cymbulopora bulloides, ii. 168.
Cymonomus quadratus, ii. 39.
Cymopolia, ii. 39.
Cymopolus asper, ii. 39.
Cypris, ii. 51.

Dactylocalyx pumiceus, ii. 172.
Dactylometra, i. 203.
Dall, W. H., on antique character of deep-sea fauna, ii. 20.
　on deep-sea mollusks and tertiary types, ii. 20.
　on gasteropods and lamellibranchs of the "Blake," ii. 62.
　Report on Mollusks, i. xxi.
Dana, J. D., i. xxi.
　on limit of reef-building corals, i. 74.
Danielssen, i. 44.
Daphnia, i. 171.
Darjiling, i. 106.
Darwin, i. 180.
　on elevation of South American coast, i. 129.
　on formation of coral reefs, i. 76.
　on limit of reef-building corals, i 74.
　on pelagic algæ, i. 208.
　on resemblance of barrier reefs and atolls, i. 72.
　theory of coral reefs, i. 55, 80.
Dasygorgia Agassizii, ii. 143.
Dawson, on climate of arctic regions, i. 107.

Davis, i. 16.
Dayman, "Cyclops" Expedition, i. 45.
De Bary on Symbiosis, i. 214.
Deep-sea acalephs, i. 186.
Deep-sea animals, carnivorous, ii. 1.
　color of, i. 310.
　habits of, i. 274.
　kept alive by ice, ii. 1.
　killed by coming to surface, ii. 1.
　looseness of their tissues, ii. 2.
Deep-sea annelids, characteristic, ii. 56.
Deep-sea beds, Fuchs on tertiary, i. 145.
Deep-sea cephalopods, i. 144.
Deep-sea corals, bathymetrical range of, i. 169.
　identity of with cainozoic, i. 162.
Deep-sea deposits, i. 143.
　Fuchs on, i. 142.
　names of, i. 263.
　of past ages, i. 141.
Deep-sea fauna, i. 153, 162.
　composition of, i. 162.
　in track of oceanic currents, i. 167.
　uniform composition of, i. 156.
Deep-sea fauna and distribution of food, i. 266.
Deep-sea fishes, ii. 21.
　color of, i. 311.
　peculiarities of, ii. 21.
　specialization of, ii. 33.
　young of, pelagic, i. 185.
Deep-sea flora, i. 166.
Deep-sea formations, i. 140.
　facies of, i. 142.
Deep-sea forms, range of, i. 302.
Deep-sea gasteropods, blind, i. 165.
Deep-sea life, physiology of, i. 294.
Deep-sea sharks, i. 40.
Deep-sea species retaining shallow-water habits, i. 166.
Deep-sea sounding, deepest by Belknap, i. 47.
　early, i. 47.
　by cup by Sands, i. 47.
　by detacher by Brooke, i. 47.
　by time intervals by W. R. Rogers, i. 47.
　with cod-line by Platt, i. 47.
　with "Hydra" machine, i. 47.
　with wire by Barnett, i. 47.
　with wire by Belknap, i. 47.
　with wire by Thomson, i. 47.
　with wire by Walsh, i. 47.
　with wire by Wilkes, i. 47.
Deep-sea sounding and dredging, Sigsbee on, i. 51.
Deep-sea temperatures, by "Challenger," i. 46.

202 INDEX.

Deep-sea temperatures, by "Gazelle," i. 46.
  by Miller-Casella thermometer, i. 46.
  by "Tuscarora," i. 46.
  Franklin on, i. 46.
  Humboldt on, i. 46.
  Lenz on, i. 46.
  Parry on, i. 46.
  Phipps on, i. 46.
  Pouillet on, i. 46.
Deep-sea thermometer, Tait, P. T., correction of, i. 16.
Deep-sea types, affinities of, i. 156.
  "Albatross" on distribution of, i. 152.
  characteristic, ii. 21.
  fossil representatives of, ii. 18.
  passage of, into continental and littoral zones, i. 163.
  predominant tint of, i. 311.
Deep-sea work, historical sketch of, i. 39.
  by Bache, i. 49.
  by Craven, i. 49.
  by Davis, i. 49.
  by Maffitt, i. 49.
  of Mitchell, i. 49.
  of Pourtalès, i. 49.
  of Sands, i. 49.
  of U. S. Coast Survey, i. 48.
Deep-sea work, publications on, i. 48.
  Depths of the Sea, i. 48.
  Moseley Notes on "Challenger" Expedition, i. 48.
  Narrative of "Challenger," i. 48.
  Thalassa by Wild, i. 48.
  Voyage of the "Challenger," i. 48.
Deep soundings by Bartlett, i. 48.
  by Brownson, i. 47, 48.
Deep-water collections, ii. 8.
Deep-water deposits, Agassiz, L., on, i. 40.
Deep-water fauna, i. 44.
Deep-water gasteropods, blind, i. 308.
Deep-water types, identity of, i. 152.
  origin of, in palæozoic times, i. 151.
Deima Blakei, ii. 86.
Deimatidæ, ii. 86.
De la Beche, on siliceous matter in water, i. 150.
Delesse lithologie, i. 49.
Deltocyathus italicus, ii. 140.
Denmark Strait, i. 242.
Dentalium perlongum, ii. 67.
Dentalium, lives in ooze, ii. 65.
  bathymetrical range of, i. 169.
Denudation, absence of aerial, i. 104.
  extent of aerial, i. 127.
  of Mississippi, Humphreys and Abbott, on, i. 128.

Deposition of sedimentary rocks, i. 130.
Deposits, affected by depth, i. 266.
  along continental coasts, i. 263.
  of past periods, compared to those of to-day, i. 167.
Depth, greatest reached by "Blake," i. 97.
Desmophyllum crista-galli, ii. 151.
Desmophyllum Riisei, ii. 150, 151.
Desmophyllum solidum, ii. 150, 151.
Diadematidæ, i. 82, 159.
Diaseris crispa, ii. 153.
Diastopora repens, ii. 78, 79.
Diatom ooze, i. 266.
Diatoms, siliceous remains of, i. 271.
Dicranodromia, ii. 40.
Dietz on Florida, i. 68.
Diopatra, bathymetrical range of, ii. 56.
Diopatra Eschrichtii, ii. 53.
Diopatra glutinatrix, ii. 53, 56.
Diopatra Pourtalesii, ii. 55.
Diphyes acuminata, ii. 135.
Discina atlantica, ii. 77.
Discophorous medusæ, ii. 130.
Discorbina orbicularis, ii. 168.
Distichopora foliacea, ii. 140.
Distribution, of animals and plants by currents, i. 117.
  of species, i. 168.
Dittmar on absorbed gases of bottom water, i. 297.
  on solids in ocean water, i. 129.
Dodecabostrycha dubia, ii. 130, 131.
Doliolum, i. 187.
Dolphin ridge, i. 123, 164.
Dolphin Rise, i. 97, 242.
Dominica, island of, i. xvii.
Dorippidoidea, ii. 38.
Dorocidaris, i. 158.
Dorocidaris Blakei, ii. 80.
Dorocidaris papillata, ii. 88.
Dredge of Ball, i. 48.
  of Edward Forbes, i. 48.
  of O. F. Müller, i. 48.
  modification of, by U. S. Coast Survey, i. 20.
  of the Philippines islanders, i. 25.
  speed in lowering, i. 29.
  used by Péron, i. 48.
  used by Stimpson, i. 48.
  used by Wilkes' Expedition, i. 48.
Dredge and trawl, sifting of contents of, i. 26.
Dredge and swabs, i. 24.
Dredging lines, 1877-78, i. ix.
  1878-79, i. xi.
  1880, i. xix.
  Tampa Bay to Mississippi, i. ix.

INDEX. 203

Dredging operations, 1877-78, i. viii.
   1878-79, i. xi.
   1880, i. xix.
Dredging in tradewinds, i. xii.
Drift of New England coast, Agassiz on, i. 122.
Driftwood, transportation of, i. 180.
Dromidæ, ii. 40.
Duncania barbadensis, ii. 155.
Duncan on fossil corals of West Indies, i. 161.
   on parasitic fungus, i. 166.
   on tertiary corals, ii. 18.
   on West Indian tertiary corals, ii. 19.
Dyplophysa, i. 181.

Earth's crust, Saporta on formation of, i. 140.
Eastern Caribbean, topography of, i. 100.
Eastern coast of North American continent, topography of, i. 93.
Echinarachnius in deep water, ii. 97.
Echini, American genera of, i. 159.
   bathymetrical range of, i. 108.
   characteristic West Indian fossil, ii. 97.
   color of deep-sea, i. 311.
   embryonic character of apical system of, ii. 91.
   list of dead tests of, ii. 97.
Echinid fauna of West Indies, analysis of, i. 159.
   former distribution of, i. 157.
   origin of, i. 157.
Echinocucumis typica, ii. 85.
Echinocyamus pusillus, ii. 97.
Echinoderms, ii. 84.
   agency of in triturating sand, i. 85.
   known previous to "Blake" Expedition, ii. 4.
Echinothuriæ, ii. 94.
   flexible test of, ii. 94.
Ehlers, Ernst, on annelids, ii. 52.
   on "Porcupine" annelids, ii. 57.
   Report on Annelids, i. xxi.
Ehrenberg on foraminifera at great depths, i. 45.
   on globigerinæ, i. 146.
   on limit of reef-building corals, i. 74.
Elasipoda, abyssal types of, ii. 84.
   huge species of, ii. 84.
Electrical thermometer of C. W. Siemens, i. 17.
Eledone verrucosa, ii. 60.
Elevated reefs of Barbados, i. 79.
   of Cuba, i. 79.
   of San Domingo, i. 79.
Elevation, lines of, in later geological periods, i. 132.

Ellis, i. 39.
   figures of Umbellula, ii. 142.
Elpididæ, ii. 86.
Enclosed basins, i. 245.
Enclosed seas, i. 245.
   density of, i. 300.
Enclosed seas and inland seas, fauna of, i. 40.
Entz, Geza, on parasitic algæ, i. 214.
Epibulia, i. 181; ii. 135.
Epimeria loricata, ii. 49.
Epizoanthus, ii. 148.
   associated with Catapagurus, ii. 41.
Epizoanthus houses, ii. 41.
Equatorial circulation, i. 132.
Equatorial current, i. 92.
   former course of, i. 113.
   hot water from, i. 233.
Equatorial drift, i. 257.
Equatorial water, path of, i. 255.
Equipment of "Blake," i. 1.
Eryonidæ, ii. 42.
   eyes of, ii. 42.
   eyes of fossil, ii. 43.
Eryon-like crustacea, i. 144.
Escharipora stellata, ii. 80.
Eucharis multicornis, ii. 130.
Eucheilota, i. 183.
Eucope, i. 159.
Eudoxia, i. 181.
Eunice conglomerans, ii. 55.
Eunice tibiana, ii. 55.
Eunicidæ, fossil, ii. 56.
   genera of, ii. 55.
   importance of, ii. 55.
   tubes of, ii. 55.
Euphausia, i. 193.
Euphronides cornuta, ii. 87.
Euplectella, Jovis, ii. 172, 173.
Eurypharynx, ii. 35.
Euthna, i. 180.
Everglades, i. 69.
   concentric reefs of, i. 63.
Extracrinus, ii. 117.
Eyes, as spectroscopes, i. 309.
   of deep-water animals, i. 307.

Faugophiliua submersa, ii. 177.
Farciminaria delicatissima, ii. 78.
Farlow, i. 209.
Farrea facunda, ii. 171.
Fauna, change of, due to nature of bottom, i. 285.
   characteristic of restricted regions, ii. 13.
   evidence furnished by fossils on character of, i. 275.
   in closed seas, isolation of, i. 104.

Fauna, of abyssal region, ii. 14.
of bottom between Windward Islands, ii. 14.
of calcareous ooze, i. 143.
of continental region, ii. 14.
of early geological periods, i. 154.
of great limestone banks, ii. 13.
of limestone plateaux, i. 143.
of littoral region, ii. 14.
of past periods uniform at great depths, i. 108.
of plateaux, i. 92.
of plateaux in track of currents, i. 92.
of Pourtalès Plateau, ii. 13.
of Pourtalès Plateau, extension of, i. 287.
of pteropod ooze, ii. 13.
of Red Sea and Mediterranean, difference of, i. 123.
of reef region, ii. 14.
of steep slope of Gulf Stream, i. 120.
of successive reefs, i. 161.
off the Tortugas, ii. 14.
Faunal districts, i. 302.
narrow limits of, i. 167.
Favosites, supposed species of, ii. 83.
Ferrell on swinging of cold water at bottom, i. 249.
Fewkes, J. Walter, i. x.
on "Blake" acalephs, ii. 120.
Report on Acalephs, i. xxi.
Fierasfer and holothurians, association of, i. 215.
Firoloiden, i. 191.
Fish Commission, U. S., explorations of, i. 50.
"Fish-Hawk," i. 50.
Fish teeth in bottom deposits, i. 281.
Fishes, bathymetrical range of, i. 108; ii. 23.
bottom feeders, ii. 27.
bottom living species, ii. 24.
cartilaginous skeletons of, i. 304.
color of deep-sea, i. 311.
known previous to "Blake" Expedition, ii. 4.
migration of, ii. 23.
of the abyssal realm, ii. 23.
pelagic types of, ii. 23.
phosphorescent light of, ii. 34.
phosphorescent organs of, ii. 36.
predaceous, ii. 27.
report on by Goode and Bean, i. xxi.
shallow-water species of, ii. 24.
tactile organs of, ii. 36.
upper limits of, ii. 23.
Flabellum Goodei, ii. 150.
Flabellum Moseleyi, ii. 150.
Flat fishes, fossil, ii. 24.

Flora, distribution of by drift, i. 122.
Florida, Agassiz on the age of, i. 88.
axis of elevation of, i. 110.
backbone of, Smith and Hilgard on the, i. 61.
bathymetrical sections off, i. 66.
bryozoa of, identical with tertiary types, ii. 79.
character of coast islands of, i. 67.
coral reef of, i. 58.
Dietz on character of, i. 68.
eastern coast-line of, i. 52.
flora of, i. 110.
geology of, Conrad on, i. 110.
geology of, by E. Hilgard, i. 110.
geology of, by E. A. Smith, i. 110.
limestone backbone of, i. 110.
limestone plateaux of, i. 122.
mangrove islands of, i. 53.
not built up by reef, i. 69.
recent limestones of, i. 62.
submarine plateau of, i. 62.
southern coast-line of, i. 52.
west bank of, i. 52.
Florida Bank, i. x.
material of, i. 56.
submarine base of, i. 141.
west edge of, i. ix.
Florida flats, keys of, i. 52.
Florida keys, structure of, i. 53.
Florida mud flats, dip of, i. 59.
Florida peninsula, age of, i. 88.
Florida plateau, animal life abundant on, i. 62.
Florida Reef, i. 52.
channels across, i. 60.
curve of, i. 57.
depth of water on, i. 54.
extension of to deeper water, i. 60.
flats of, i. 58.
mud flats, appearance of, i. 59.
northern extension of, i. 69.
not elevated, i. 61.
Pourtalès exploration of, i. 286.
recent corals of, i. 78.
shape of, i. 53.
survey of, i. vii.
survey of, by L. Agassiz, i. 49.
tides across the, i. 57.
Food question in distribution of animals, i. 91.
Food supply of young fishes, i. 204.
Foraminifera and currents, i. 279.
Foraminifera, arenaceous types of, ii. 158.
as guides to deep-sea deposits, i. 146.
at great depths, Huxley on, i. 45.
bathymetrical range of pelagic, i. 196.

INDEX.

Foraminifera, Brady on, ii. 157.
　Carpenter on, ii. 158.
　Ehrenberg on, at great depths, i. 45.
　　in littoral deposits, i. 146.
　Parker on, ii. 158.
　Williamson on, ii. 158.
Foraminiferous calcareous bottom, Pourtalès on, i. 284.
Forbes, Edward, on limit of animal life, i. 40.
　on Ægean Sea, i. 40.
Forel, on penetration of light, i. 305.
　on pelagic fauna of Swiss lakes, i. 199.
Formigas Bank, i. 98.
Fort Jefferson laboratory, i. xi.
Foster on temperature of ocean, i. 46.
Franklin, i. 247.
　on deep-sea temperatures, i. 46.
Frankland on solid matter held in solution, i. 128.
Fossil marine fauna, i. 122.
Fuchs, on bathymetrical faunal subdivisions, i. 163.
　on deep-sea deposits, i. 142.
　on tertiary deep-sea beds, i. 145.
Fungia symmetrica, ii. 153.

Gadus fossil, ii. 25.
Galatheoidea, ii. 42.
Garay, Governor of Jamaica, i. 251.
Gardner, J. S., on blue muds, i. 148.
　on temperature of bottom of ocean, i. 132.
Garman, Samuel, i. x.
Gas analysis, Behrens, apparatus for, i. 204.
　Bunsen apparatus for, i. 204.
Gases, presence of at great depth, i. 23.
Gasteropods, ii. 62.
　roving life of, ii. 65.
　small size of deep-sea, i. 109.
Gastrostomus Bairdii, ii. 31, 34, 35.
Gaza superba, ii. 68.
Geddes on Philozoön, i. 214.
Geikie, A., on great age of continents and oceanic basins, i. 127.
　on lost Atlantis, i. 127.
Gephyreans, ii. 52.
Geographical provinces, i. 302.
Geographical range of polyps, ii. 17.
Geography, of different geological periods, i. 132.
　of time of chalk, i. 133.
　of tertiary period, i. 134.
　since triassic period, i. 161.
Geological time since the cretaceous, i. 130.
George's Bank, i. 50.
George's Shoal, i. xix.

Geryon quinquedens, ii. 38.
Giant squids, ii. 62.
Gibbs, W., determination of specific gravity, i. 21.
Giglioli, i. 41.
Glaucus, i. 186.
Gleba hippopus, ii. 134.
Globigerina, i. 171.
Globigerina and pteropod ooze of Gulf of Mexico and Caribbean, i. 281.
Globigerina bottom ooze, discovery of, i. 284.
Globigerina bulloides, ii. 167, 168.
Globigerina ooze, i. 272.
　compared to chalk, i. 147.
　composition of, i. 147.
　depth at which found, i. 265.
　depth at which it appears, i. 147.
　Murray on, carbonate of lime in, i. 281.
　northern extension of, i. 282.
　transition of to red clay, i. 272.
Globigerinæ, Bailey on Atlantic, i. 146.
　bottom specimens of, ii. 167.
　living on bottom, ii. 150.
　Maffitt and Craven on Gulf Stream, i. 45.
　Müller on, ii. 150.
　Murray on, ii. 150.
　Owen on, ii. 159.
　Pourtalès on, ii. 159.
　Pourtalès on limit of, i. 284.
　shell of, ii. 167.
　young shells of, ii. 167.
Glossocodon, i. 186.
Glyphocrangon aculeatus, ii. 45.
Gnathophausia Zœa, ii. 48, 49.
Gobies, ii. 29.
Goës, i. 42.
Goniasteridæ, ii. 102.
Goniocidaris, radioles of, ii. 80.
Goniopecten, ii. 103.
Gonostoma microdon, ii. 0.
Goode and Bean, notes on deep-sea fishes, ii. 21.
　Report on Fishes, i. xxi.
Goose fish, ii. 24.
Gorgoniæ, deep-sea types of, ii. 143.
　phosphorescence of, i. 199; ii. 146.
Graff, L. v., Report on Myzostomidæ, i. xxi.
　on Myzostomidæ of the "Blake," ii. 126.
Grand Cayman, great depth off, i. 100.
Grande Terre Guadeloupe, i. xvi, 63, 65.
Gravitation theory of oceanic circulation, i. 247.
Great Britain, former connection of, i. 125.
Greater West India Islands, affinity of fauna and flora of, i. 115.

Greenland current, i. 241.
Greensand, Bailey on, i. 45.
   Bailey on process of formation of, i. 278.
   deposits of, i. 141.
   Ehrenberg on, i. 278.
   modern, i. 278.
   Murray on composition of, i. 278.
   Pourtalès on formation of modern, i. 278.
   Pourtalès on position of belt of, i. 278.
   on shore edge of Gulf Stream, i. 236.
   off Cayo de Moa, i. 201.
Grenada, island of, i. xviii.
Grenadines, i. xviii.
Groupers, ii. 28.
Guadeloupe, island of, i. xvi.
Guinea Stream, i. 247.
Guppy on reef holothurians, i. 85.
Gulf of Maine, bottom specimens of, i. 261.
Gulf of Mexico, basin of, i. 100.
   a hydrostatic reservoir, i. 240.
   areas of basin of, i. 101.
   bottom temperature of, i. 15.
   heaping up of water in, i. 248.
   Howell exploration of, i. 50.
   superheated water of, i. 233.
   topographical features of, i. 102.
   unstable equilibrium of, i. 248.
Gulf Stream, i. xx, 241.
   A. Agassiz report on, i. xxi.
   amount of outflow of, i. 256.
   ancient course of, i. 137.
   Bache on, in 1845, i. 252.
   Bache on warm and cold bands of, i. 253.
   Baer, K. E. v., on eastern extension of, i. 252.
   Bailey on soundings of along course of, i. 272.
   Bartlett on course of, i. 257.
   Bartlett on warm and cold bands of, i. 254.
   Blagden on temperature of, i. 252.
   Coast Survey on structure of, i. 253.
   cold bands of, i. 234.
   course of, across Gulf of Mexico, i. 113.
   Craven on warm and cold bands of, i. 253.
   Davis on warm and cold bands of, i. 253.
   development of knowledge of, i. 240.
   eastern extension of, i. 257.
   Folger on, i. 252.
   Franklin exploration of, i. 252.
   inflow of, into Gulf of Mexico, i. 256.
   Maffitt on warm and cold bands of, i. 253.

Gulf Stream, Mitchell, H., current observations of, i. 232.
   northern course of, i. 121, 138.
   outflow of, into Straits of Florida, i. 256.
   path of, towards Europe, i. 257.
   temperature of, near shore, i. 258.
   Thomson on cold and warm belts of, i. 254.
   trough of, i. xx.
   velocity of, i. 256.
   velocity of axis of, i. 259.
   velocity of, through Yucatan Channel, i. 258.
   wearing action of, i. 138.
Gulf Stream fauna off Charleston, i. 120.
Gulf Stream floor swept clean, i. 236.
Gulf Stream slope, i. 90, 274.
Gulf Stream work, i. xi.
Günther on accessory eyes of fishes, ii. 22.
   on deep-sea fishes, ii. 21.
Gurnards, ii. 30.
Guyot on great age of continents and oceanic basins, i. 127.

Habits of deep-sea animals, ii. 8.
Haeckel, E., i. 35.
   on Bathybius, i. 204.
   on yellow cells of radiolarians, i. 213.
Hake, ii. 23.
Halcyonoids, ii. 142.
Halimeda, i. 82.
Halobates, i. 179.
Halosaurus macrochir, ii. 32, 33.
Hand nets, collecting by, i. 35.
Haplophyllia paradoxa, ii. 154, 155.
Harger, Report on Isopods, i. xxi.
Harvey, i. 313.
"Hassler" Expedition, i. 40, 51.
Hastigerina, i. 196; ii. 108.
Hatteras, lines run normal to coast south of, i. 135.
   slope of detritus off, i. 131.
Hébert on aragonite, i. 147.
Helioporidæ, ii. 138.
Hemiaster expergitus, ii. 100.
Hemiaster fascioles of, ii. 98.
Hemiaster zonatus, ii. 100.
Hemieuryale pustulata, ii. 5.
Hemipedina, i. 158.
Hemipedina cubensis, ii. 97.
Hermit crabs, ii. 40.
Herschell, i. 247.
Hertwig on yellow cells of actiniæ, i. 214.
Heterocarpus carinatus, ii. 40.
Heteropods, i. 190.
Heteropora, ii. 83.

Hexactinellidæ, ii. 12, 170.
Thomson, C. Wyville, on, ii. 170.
Hierlatz, liassic beds of, i. 144.
Hilgard Deep, i. 100.
Hilgard, E., geology of Florida, i. 110.
Hilgard, J. E., i. 49, 93.
   on Gulf of Mexico as a hydrostatic reservoir, i. 249.
Hilgard's aerometer, i. 21.
Hippothoa biaperta, ii. 81.
Hippurella annulata, ii. 137, 138.
Histriophorus, i. 193.
Hoek on palæontology of cirripeds, ii. 51.
Hoisting engines for dredge, i. 31.
Holopus, ii. 6.
   young, ii. 124.
Holopus Rangi, ii. 123.
Holothurians, ii. 84.
   known previous to "Blake" Expedition, ii. 4.
   shallow-water genera of, ii. 84.
Holtenia, i. 40.
Holtenia Pourtalesii, ii. 175.
Homalodromia, ii. 40.
Homolidæ, ii. 40.
Homolopsis, ii. 40.
Hooker, i. 39.
   on giant kelp, i. 209.
Humboldt, i. 247.
Humphreys and Abbott on Mississippi denudation, i. 128.
Hunt, E. B., on mud flats of Florida, i. 60.
   theory of coral reef of Florida, i. 55.
Hutton on New Zealand fauna, i. 122.
Huxley, Th. H., on foraminifera at great depths, i. 45.
Hyalea, i. 187, 265.
Hyalinœcia tubicola, ii. 52.
Hyalinœcia tubes, ii. 57.
Hyalonema boreale, ii. 177.
Hyalonema, Japanese, ii. 173, 174, 175.
   Leidy on siliceous spicules of, ii. 173.
Hyalonema Sieboldii, ii. 173.
Hyalopomatus Langerhansi, ii. 53, 57.
Hydrocorallinæ, Moseley on, ii. 138.
Hydroids, report on by George J. Allman, i. xxi.
   Allman on deep-sea, ii. 135.
   in fresh water, i. 153.
   known previous to "Blake" Expedition, ii. 7.
   S. F. Clark on, i. xxi.
Hydromedusæ, ii. 128.
Hymenodiscus Agassizii, ii. 105, 106.
Hyperammina elongata, ii. 102.
Hypsicometes, ii. 30.

Ice-borne deposits and rocks, i. 271.
Infulasteridæ, i. 159.
"Ingegerd" and "Gladan," Expedition of the, i. 40.
Iguana of Navassa, i. 115.
Inland seas and enclosed seas, fauna of, i. 40.
Inshore plateau, deposits on, i. 201.
Intermediate deep-sea forms, ii. 17.
Intermediate depths, "Challenger" Expedition on specimens from, i. 200.
Invertebrates, coloring matter of, i. 309.
   subdivision of labor among, i. 215.
Ipnops, eyes of, ii. 34.
Ipnops Murrayi, ii. 32.
Iridogorgia Pourtalesii, ii. 144.
Islands forming Central and South America, i. 113.
Isocardia cor, ii. 74.
Isolated rocky patches in Gulf of Mexico, ii. 57.
Isopod, blind, ii. 178.
   largest known, ii. 49.
Isopods, ii. 48.
   report on by Harger, i. xxi.
Isospondyli, pelagic, ii. 33.
Isthmus of Tehuantepec, closing of passage across, i. 118.

Jacoby, H. M., i. viii, 32.
Janthina, i. 186.
Jeffreys, Gwyn, i. 43.
   on depth of cretaceous sea, i. 146.
Jenkin, Fleeming, report on animals attached to submarine cable, i. 44.
Johnson, J. A., i. 49.
"Josephine," expedition of the, i. 42.
Julien, A. A., i. 115.

Kelp, Hooker on giant, i. 209.
Keys, formation of, i. 54.
Keys formed by waste, i. 57.
Key West Harbor, entrance to, i. 54.
Key West, Navy Depot, i. xi.
Kinchinjinga, i. 106.
"Knight Errant," Expedition of the, i. 44.
Kohl, Geschichte des Golf Strom, i. 250.
Kölliker on "Challenger" pennatulids, ii. 143.
Kophobelemnon scabrum, ii. 142.
Krümmel on elevation of continental masses, i. 126.
   on oceanic basins, i. 126.
Krusenstern, temperatures taken by, i. 46.
Kurtz, i. 43.

Labrador current, i. 241.
Lagena distoma, ii. 165.
Lamellibranchs, ii. 62.
Land animals, distribution of, from the arctic, i. 100.
Land connections, former, i. 121.
Land tortoise of West Indies, i. 115.
Langley on color of atmosphere, i. 307.
Larval forms, retardation of development of, i. 175.
Leconte, Joseph, theory of coral reef of Florida, i. 55.
Lefroyella decora, ii. 171.
Lemuria, i. 126.
Lenz, i. 249.
 on deep-sea temperatures, i. 46.
Lepas anatifa, i. 182.
Lepidisis, ii. 145.
Leptocephalus, i. 121.
 a larval form, i. 175.
Leptonemus discus, ii. 154.
Leptothyra induta, ii. 68, 69.
Lescarbot on warm water in North Atlantic, i. 252.
Leucosoidea, ii. 38.
Level, difference of, between Sandy Hook and Mississippi, i. 249.
Leydig, on accessory eyes of fishes, ii. 22.
 on phosphorescent organs of fishes, ii. 22.
Light, absence of, beyond 100 fathoms, i. 165.
 penetration of, i. 305.
 Pourtalès on penetration of, i. 305.
 Sarasin and Fol on penetration of, i. 305, 306.
 Sarasin and Soret on penetration of, i. 306.
 Verrill on color of at great depths, i. 305.
"Lightning," deep-sea temperatures by, i. 46.
 Expedition of the, i. 43.
Limestone banks, of Gulf of Mexico, i. 63.
 submarine position of, i. 65.
Limestone and volcanic peaks, i. 64.
Limestone, formed by pelagic animals, i. 85.
 deposits of coarse, i. 141.
 of Caribbean, i. 63.
 of Grande Terre, i. 63.
 of Southern Cuba, i. 63.
 of West India Islands, i. 64.
 of Yucatan and other plateaux, i. 72.
Limestone deposits, formation of, i. 90.
Limestone plateaux, i. 131.
Limestone terraces, of Windward Islands, i. 64.
 of Barbados, i. 63.

Limopsis aurita, ii. 73.
Limit of animal life, Edward Forbes on, i. 40.
Limits of "Blake" dredgings, ii. 12.
Line of soundings, across eastern Caribbean, i. 100.
 east of Cape May, i. xx.
 from Cape San Antonio to Sand Key, i. ix.
 from Curaçoa to Alta Vela, i. 100.
 from Curaçoa to mainland, i. 99.
 from Yucatan Bank to Alacran Reef, i. ix.
 north of Cape Hatteras, i. xx.
Lindahl on Umbellula, i. 40.
Lindenkohl, on 140 fathom hole, i. 95.
Linerges mercurius, i. 186.
Lingula, a littoral species, i. 160.
 a shallow-water animal, i. 151.
Liparidæ, ii. 28.
Liriope, i. 183.
List of foraminifera found in globigerina ooze, i. 265.
Lithistidæ, ii. 12, 175.
 Zittel on, ii. 175.
Lithodes Agassizii, ii. 39, 40, 41.
Littoral belt, i. 157.
Littoral bottom deposits, i. 260.
Littoral fauna, i. 162, 165.
 position of, i. 141.
Littoral regions, character of fauna of, i. 107.
Littoral species, bathymetrical range of, i. 302.
Littoral types, adaptation of, i. 207.
 migration of, i. 207.
Lituolinæ, ii. 163.
Liversidge, i. 148.
Ljungman, i. 42.
Loggerhead Key, oölitic and breccia limestone at, i. 67.
Lophius piscatorius, ii. 31.
Lophohelia prolifera, ii. 151.
Lopholatilus chamæleonticeps, ii. 29.
Loriol, P. de, ii. 99.
Lovén, S., i. 42.
 on derivation of abyssal fauna, i. 155.
 on geographical range of deep-sea types, ii. 15.
Lucifuga of caves of Cuba, ii. 26.
Lump fishes, ii. 28.
Lump suckers, ii. 28.
Lütken, on West Indian Pentacrinidæ, ii. 118.
Lycodidæ, ii. 26.
Lyell, on land and oceanic hemispheres, i. 241.
Lyman, T., Report on Ophiurans, i. xxi.

Macrocystis, i. 313.
Macropneustes, i. 159, 160.
Macropneustes spatangoides, ii. 98.
Macruroids, ii. 26.
Macrurus Bairdii, ii. 26, 27.
Macrurus caribbæus, ii. 26, 27.
Madagascar, relation to Africa, i. 126.
Madrepora cervicornis, i. 82, 86.
Madrepora palmata, i. 70, 83.
  wall of, i. 70.
Madrepora prolifera, i. 81, 82.
Mæandrina areolata, i. 82.
Maffitt and Craven on Gulf Stream globigerinæ, i. 45.
Magnaghi, i. 41.
Maioidea, ii. 37.
Malacosteus niger, ii. 34, 35.
Maldane cuculigera, ii. 54.
Mallet on primordial ocean, i. 153.
  on atmosphere of steam, i. 128.
Malmgren, i. 42.
Malthe, ii. 31.
Mammals, distribution of, by driftwood, i. 122.
Manatee bones, i. 282.
  Pourtalès on, i. 144.
Manganese concretions and phosphate of lime, i. 275.
Manganese nodules, i. 141, 290.
Mangrove islands, formation of, i. 53.
Mangrove plants, i. 53.
Manicina, i. 55.
Margarita regleës, ii. 68.
Marginella succinea, ii. 60.
Marginella Watsoni, ii. 69, 70.
Marie-Galante, i. xvi.
Marine animals, destruction of, i. 274.
  food of, i. 204.
Marine deposits, character of, i. 266.
  composition of, i. 262.
Marine fishes in the Amazons, i. 153.
Marine forms within 100 fathom line, i. 142.
Marine plants, distribution of, i. 312.
Marquesas atoll, structure of, i. 73.
Marquesas islands, examination of, i. 73.
  lagoon of, i. 73.
Marshall, his tangle bar, i. 26.
Martinique, i. 112.
  island of, i. xvi.
Mascarene islands, relation to Africa, i. 126.
Mastigoteuthis Agassizii, ii. 59.
McClintock, i. 45.
McRea, Henry, i. 32.
Mediterranean, contrast between northern and southern shores of, i. 123.

Mediterranean, great changes in eastern basin of, i. 123.
  recent connection with Atlantic, i. 123.
Medusæ, decomposition of, ii. 128.
  deep-sea types of, ii. 128.
  of intermediate depths, ii. 128.
Meiocardia Agassizii, ii. 74.
Melanocetus, ii. 32.
Mellita, i. 150.
Membranipora canariensis, ii. 70.
Meningodora, ii. 47.
Mentz, Geo. W., i. viii, 32.
Meoma, i. 150.
Merlucius, ii. 25.
Metacrinus, ii. 116.
Metacrinus angulatus, ii. 117.
Metamorphoses of deep-sea animals, ii. 8.
Mexico, Gulf of, Hydrographic Chart of, i. 15.
  unstable equilibrium of, i. 248.
Micropanope pugilator, ii. 38.
Migration of sharks, i. 123.
Miliolinæ, i. 271.
Millepora, Agassiz on, ii. 138.
  Moseley on, ii. 138.
Millepora alcicornis, ii. 138, 139.
Miller on Pentacrinus, ii. 116.
Miller-Casella thermometer, i. 15.
Milne-Edwards, Alph., i. 40, 44, 204.
  on fauna of antarctic region, i. 121.
  Report on Crustacea, i. xxi.
  on Sargasseum, i. 212.
Milne-Edwards and Haime on tertiary corals, ii. 18.
Mississippi basin, denudation of, i. 128.
Mississippi mud, i. 57, 131.
  fauna of, i. 282.
Mississippi, wearing action of, i. 138.
Mitchell, Henry, on current of Gulf Stream, i. 232.
  on Nicholas and Santaren channels, i. 235.
Mitra Swainsonii, ii. 70.
Mixtopagurus paradoxus, ii. 41.
Mnemiopsis, i. 199.
Modiola polita, ii. 64, 73.
Mohn, i. 44.
Mollusk fauna of West Indies, i. 114.
Mollusks, ii. 58.
  absence of vegetable feeders, ii. 64.
  bathymetrical range of, i. 109.
  carnivorous, ii. 65.
  Dall, W. H., Report on, i. xxi.
  distribution of, by driftwood, i. 122.
  flexible species of, ii. 65.
  inflexible species of, ii. 65.

Mollusks, known previous to "Blake" Expedition, ii. 4.
  with poison fangs, ii. 65.
Mona Passage, i. 98, 159.
Monactinellidæ, ii. 177.
Monolene atrimana, ii. 24.
Montauk Point, i. xix.
Montserrat, island of, i. xvi.
Moore, W. S., i. x, 32.
Moseley, on absence of palæozoic types in deep sea, i. 157.
  on age of abyssal fauna, i. 155.
  on coloring matter of deep-sea invertebrates, i. 300.
  on derivation of abyssal fauna, i. 155.
  on eyes of Ipnops, ii. 34.
  on fauna and flora of deep-sea explorations, i. 156.
  on fructification of Sargassum, i. 212.
  on pelagic conditions of littoral types, i. 154.
  on range of temperature, i. 300.
  on sinking of Salpæ, i. 187, 313.
Mosquito Plateau, i. 98.
Mount Maitland, i. xviii.
Mud flats of Florida, E. B. Hunt on, i. 60.
Mud holes off New York, i. 272.
Müller, Johannes, i. 35, 200.
  on yellow cells of radiolarians, i. 213.
Müller and Troschel, on number of West Indian ophiurans, ii. 113.
Müller, O. F., dredge of, i. 24.
Mülleria, i. 82.
Munida, ii. 43.
Munidopsis rostrata, ii. 42, 43.
Murray, John, i. 4, 44.
  on carbonate of lime in sea water, i. 65.
  on chondres and cosmic dust, i. 262.
  on Coast Survey bottom deposits, i. 280.
  on composition of red clay, i. 267.
  on depth of chalk sea, i. 147.
  on floating pumice stone, i. 267.
  on material held in suspension, i. 80.
  on shallow-water deposits, i. 280.
  on solvent action of sea water, i. 65.
  on Tahiti Reef, i. 77, 88.
  on typical bottom deposits of Caribbean, i. 288.
  on use of tow-net in deep water, i. 202.
  Report on submarine deposits, i. xxi.
Murray and Abbé Renard on bottom deposits, i. 261.
Murray and Pourtalès, on bottom deposits south of Cape Hatteras, i. 275.
Mysis, i. 179, 193.
Myxine glutinosa, ii. 36.

Myzostoma Agassizii, ii. 127.
Myzostoma cysticolum, ii. 127.
Myzostoma filicauda, ii. 127.
Myzostomidæ, Report on, by L. v. Graff, i. xxi.

Nanomya, i. 181.
Nares, i. 44.
Nares Deep, i. 106.
Nebalia, i. 193.
Nectoteuthis Pourtalesii, ii. 50.
Nematocarcinus cursor, ii. 46.
Nematocarcinus ensiferus, ii. 46, 47.
Nemertinæ, ii. 52.
Nemichthys scolopaceus, ii. 34, 35.
Neohela pasma, ii. 49.
Neolampas rostellata, ii. 97, 98.
Nephropsis Agassizii, ii. 43, 44.
Nettastoma procerum, ii. 34, 35.
Nevis, island of, i. xvi.
New England coast, Agassiz on wearing of, i. 122.
  cold belt along, i. 119.
  warm belt along, i. 119.
New Zealand, relation to Australia, i. 125.
New Zealand fauna, Hutton on, i. 122.
Nicholas and Santaren channels, Henry Mitchell on, i. 235.
Noctiluca, i. 196.
  phosphorescence of, i. 196.
Nodosaria, list of varieties of, ii. 166.
Nodosaria communis, ii. 166.
Nodosaria radicula, ii. 166.
Nordenskiöld, i. 42.
Norman, A. M., i. 43.
  on deep-sea North Atlantic species, i. 102.
Norse navigators and Labrador current, i. 250.
North America, at time of chalk, i. 133.
  archæan continent of, i. 129.
North Atlantic, area of maximum temperature in, i. 243.
  isolation of, i. 243.
Notacanthus phasganorus, ii. 30.
Notostomus, ii. 47.
Nourse, C. J., i. x, 32.
Norway haddock, ii. 24.
Nucleolidæ, ii. 97.
Nullipores, masses of, i. 82.
Nummulinidæ, ii. 169.
Nymphon, ii. 50.

Ocean water, Boguslawski on solids in, i. 129.
  density of at equator, i. 248.
  Dittmar on solids in, i. 129.
  solvent power of, i. 147.

Oceanic basins, Agassiz, L., on, i. 4.
  Krümmel on, i. 126.
  permanence of, i. 125.
  pressure on rocks below, i. 132.
  soundings in, by "Challenger," i. 260.
  temperature of, i. 246.
  Thomson, C. Wyville, on, i. 4.
  topography of, i. 107.
Oceanic circulation, tradewind theory of, i. 247.
  Thomson's theory of, i. 247.
Oceanic currents, theories of, i. 247.
  in past ages, i. 128.
  slow movements of, i. 302.
Oceanic deposits, organic ooze and red clay of, i. 264.
Oceanic districts, salinity of, i. 248.
Oceanic islands, i. 117.
Oceanic realms, specialization of, i. 160.
Oceanic temperature, disturbing factors of, i. 248.
Ocyroë cristallina, ii. 129.
Ocyroë, Fewkes, J. W., on, ii. 129.
Ocyroö maculata, ii. 129.
Oersted on bathymetrical belts, i. 102.
Old Bahama Channel, i. 2.
Old-fashioned types in shallow water, i. 156.
Oölitic and breccia limestone at Loggerhead Key, i. 76.
Oölitic limestone, modern, i. 286.
Ooze adapted for preservation of animals, i. 170.
Ophiactis swarming on sponges, ii. 113.
Ophidiidæ, ii. 56.
Ophidium cervinum, ii. 26.
Ophiernus, ii. 5.
Ophiocamax hystrix, ii. 5, 110, 111, 114.
Ophioconis miliaria, ii. 111, 112.
Ophiocreas, ii. 5.
Ophiocreas spinulosus, ii. 100, 114.
Ophiohelus umbella, ii. 116.
Ophiolipus Agassizii, ii. 115.
Ophiomastus secundus, ii. 113.
Ophiomitra valida, ii. 115.
Ophiomusium Lymani, ii. 114.
Ophiomusium planum, ii. 111, 112.
Ophiomyces frutectosus, ii. 111, 113.
Ophiomyxa flaccida, ii. 113.
Ophiopæpale Goësiana, ii. 111.
Ophiophyllum petilum, ii. 110.
Ophiothrix, colonies of, ii. 113.
Ophiozona nivea, ii. 5, 110.
Ophiura Elaps, ii. 111.
Ophiurans, ii. 109.
  bathymetrical range of, ii. 114.

Ophiurans, known previous to "Blake" Expedition, ii. 5.
  phosphorescence of, i. 190.
  Report on, by T. Lyman, i. xxi.
Ophiuridæ, Lyman on, ii. 100.
Opisthoteuthis Agassizii, ii. 58.
Oplophorus, ii. 47.
Orbiculina adunca, ii. 160.
Orbitolites, ii. 100, 161.
Orbulina, i. 194.
Orbulina universa, ii. 166, 167.
  Krohn on, ii. 167.
  Pourtalès on, ii. 167.
Organic matter, at distance from shore, limited supply of, i. 269.
  as food for deep-sea life, i. 313.
  held in suspension near shore, i. 269.
Organs of sense, in deep-sea fishes, ii. 22.
  great development of, in embryos, i. 176.
Organs of vision of deep-sea invertebrates, i. 165.
Orthagoriscus, i. 193.
Oscillations of earth's surface, i. 126.
Ostracods, ii. 51.
Ostraconotus spatulipes, ii. 42.
Otoliths of fishes, in bottom deposits, i. 281.
  in fine muds, i. 145.
Otter, von, i. 42, 130; ii. 142.
Oxygen and carbonic acid in sea water, i. 297.
Oxygen in sea water, Dittmar on, i. 295.
  Jacobsen on, i. 295.

Pacific and Atlantic isotherms, i. 248.
Pælopatides confundens, ii. 88.
Paguroidea, ii. 40.
Palæchinidæ, ii. 94.
Palæotropus Josephinæ, ii. 100.
Paleopneustes hystrix, ii. 100.
Paliuurus, i. 175.
Pallenopsis, ii. 50.
Panceri on phosphorescence of marine animals, i. 198.
Pandalus, ii. 46.
Paracyathus confertus, ii. 149, 150.
Parasitic algæ, Geza Entz on, i. 214.
Parasitic fungus, P. M. Duncan on, i. 166.
Parasitism, different kinds of, i. 215.
Parry on deep-sea temperatures, i. 46.
Passage of littoral to abyssal regions, ii. 7.
Passages between Windward Islands swept clean, i. 230.
Patterson, Carlile P., i. vii, 49.
Patterson Deep, i. 106.
Pecten Dalli, ii. 72.
Pecten phrygium, ii. 72.
Pecten Pourtalesianum, ii. 73.

Pedata, ii. 84.
Pediculati, ii. 30.
Pedro Bank, formation of, i. 69.
Peirce, Benjamin, i. 49, 122.
Pelagic algæ, i. 208.
　Darwin on, i. 208.
Pelagic animals, at great depths, i. 200.
　carcasses of, on bottom, i. 313.
　colors of, i. 171.
　distribution of, i. 160.
　great destruction of, i. 205.
　habitat of, i. 177.
　living at intermediate depths, i. 185.
　method of collecting, i. 179.
　phosphorescence of, i. 171.
　protective agencies to, i. 205.
　range in depth of, i. 177.
　range of, i. 306.
　struggle for food of, i. 205.
　struggle for existence in, i. 204.
Pelagic annelids, i. 193.
Pelagic cephalopods, i. 191.
Pelagic crustacea, i. 193.
Pelagic deposits, i. 143.
Pelagic embryos, i. 175.
Pelagic fauna, i. 34, 121.
　"Challenger" naturalists on the, i. 35.
　Moseley on the origin of the, i. 208.
　of eastern Caribbean, i. 198.
　of Gulf of Mexico, i. 198.
　of Gulf Stream, i. 203.
　of Swiss Lake, Forel on, i. 199.
　origin of, i. 208.
　sinking of, i. 202.
Pelagic fauna and flora, i. 171.
　sinking of, i. 203.
Pelagic fishes, ii. 9.
　large size of some, i. 103.
　localities of, ii. 22.
　organs of sense of, i. 176.
　phosphorescence of, i. 177.
Pelagic foraminifera, Bailey on, i. 272.
Pelagic globigerinæ, i. 193.
Pelagic hemiptera, i. 179.
Pelagic larvæ, transportation of, i. 208.
Pelagic mollusks, i. 187.
Pelosina, ii. 162.
Pemberton, J. H., i. 32.
Penæidæ, ii. 47.
Pennatula aculeata, ii. 142, 143.
Pennatulæ, phosphorescence of, ii. 142.
Pentacheles sculptus, ii. 42, 43.
Pentacrinidæ, relation of, ii. 116.
Pentacrinin, i. 309.
Pentacrinoids, depths at which found, i. 144.

Pentacrinus asterius, ii. 116, 117.
Pentacrinus, bathymetrical range of, i. 169.
Pentacrinus Blakei, ii. 119.
Pentacrinus decorus, ii. 118, 119.
Pentacrinus, forest of, ii. 7.
　mode of life, ii. 118.
Pentacrinus ground, i. viii.
Pentacrinus Mülleri, ii. 118, 119.
Pentagonaster ternalis, ii. 102.
Periphylla hyacinthina, ii. 131, 132.
Peristedium longispatha, ii. 30.
Péron, i. 48.
Perrier, E., Report on Starfishes, i. xxi.
　on derivation of abyssal fauna, i. 155.
　on starfishes of the "Blake," ii. 102.
Persons, i. 32.
Peterman on Florida Stream, i. 257.
Peters, G. H., i. viii, 32.
Petromyzon marinus, ii. 36.
Phakellia tenax, ii. 177, 178.
Phascolosoma, in Dentalium, ii. 52.
Pheronema Annæ, ii. 174, 175.
Philozoön, Geddes on, i. 214.
Philozoön and Polyclonia, association of, i. 215.
Philozoön and Velella, association of, i. 215.
Phipps, deep-sea temperatures by, i. 46.
Phoberus cœcus, ii. 44, 45.
Phosphorescent animals, Carpenter, W. B., and Thomson on, i. 308.
　Moseley on, i. 308.
Phornosoma placenta, ii. 95.
Phorus, inhabited by annelids, ii. 52.
Phosphate and carbonate of lime, Murray on deposition of, i. 282.
Phosphatic concretions, Murray on, i. 281.
Phosphorescence, of marine animals, Panceri on, i. 199.
　of pelagic fishes, i. 179.
　of ophiurans, i. 199.
　Verrill on protective, i. 308.
Phronima, i. 176.
Phronima and Doliolum, association of, i. 215.
Phycis Chesteri, ii. 26, 27.
Phycis fossil, ii. 25.
Phycis regius, electric, ii. 23.
Phyllosoma, larval stage of, i. 175.
Physalia, i. 180.
Physical conditions, variations of, in contiguous areas, i. 154.
Physophores, ii. 133.
Pikermi, mammals of, i. 124.
Pilumnus, ii. 38.
Pisagua, elevation of coast near, i. 129.
Pisolambrus nitidus, ii. 37, 38.

Plagusia, i. 175.
　range of vision of, i. 177.
Plants, absent in deep water, i. 107.
Platt, R., i. xii.
Platydia anomioides, ii. 77.
Plectromus suborbitalis, ii. 28.
Pleurocarpa ramosa, ii. 137.
Pleuropus, i. 187.
Pleurotoma (Ancistrosyrinx) elegans, ii. 66.
Pleurotoma Blakeana, ii. 66.
Pleurotoma curta, ii. 66.
Pleurotoma limacina, ii. 66.
Pleurotoma subgrundifera, ii. 66.
Pleurotomaria Adansoniana, ii. 69.
Pleurotomaria, bathymetrical range of, i. 160.
Pleurotomaria Quoyana, ii. 69.
Pleurotomidæ, ii. 66.
Pliobothrus symmetricus, ii. 138, 139.
Plumularidæ, ii. 135.
Pneumodermon, i. 121.
Podocidaris sculpta, ii. 92.
Pollicipes, ii. 51.
Polycistinæ, i. 195.
Polyclonia, i. 177.
　lives on the bottom, i. 185.
Polymorphina ovata, ii. 166.
Polyps, ii. 142.
Polystomella crispa, ii. 169.
Polystomellæ, i. 271.
Polytrema miniaceum, ii. 169.
Pomalostegus stellatus, ii. 57.
"Porcupine," i. 245.
　cruise of the, i. 40.
　deep-sea temperatures by, i. 46.
　expedition of the, i. 43.
Porina subsulcata, ii. 82.
Porites clavaria, i. 82.
Porites embryo, i. 74.
Porites furcata, i. 82.
Porocidaris, i. 158.
Porocidaris Sharreri, ii. 90, 91.
Poromya, ii. 73.
Porpita, i. 121, 180.
Porpitidæ, affinities of, i. 184.
Port de France, i. xvi.
Porto Rico, land shells of, i. 115.
Pourtalès, L. F., i. vii, 3.
　exploration of, in "Bibb," i. 45.
　exploration of, in "Corwin," i. 45.
　on affinity of deep-sea corals and tertiary fossils, ii. 17.
　on boundary of siliceous sand, i. 279.
　on characteristic foraminifera, i. 271.
　on corals of the "Blake," ii. 148.
　on corals of West Indies, ii. 18.

Pourtalès, L. F., on deep-sea corals and tertiary types, ii. 19.
　on penetration of light, i. 305.
　Report on Corals, i. xxi.
Pourtalès Deep, i. 100.
Pourtalès Plateau, i. 62, 284, 286.
　fauna of, i. 91.
Pourtalesia miranda, ii. 101.
Pourtalesiæ, i. 159.
　affinities of, ii. 101.
　forerunners of spatangoids, ii. 97.
Prayа, ii. 135.
Pre-archæan continent, i. 127.
Preservation of animals in bottom deposits, i. 170.
Pressure, adaptation of marine animals to, i. 304.
　range of, for marine animals, i. 301.
Primnoa Pourtalesii, ii. 146.
Primordial ocean, fauna of, i. 153.
　Mallet on, i. 153.
Protoplasm, i. 149.
Protozoans without solid tests, ii. 157.
Pseudodiadematidæ, i. 159.
Psolus tuberculosus, ii. 85.
Psychropotes longicauda, ii. 86.
Pterophryne, nest of, ii. 31.
Pterophysa grandis, i. 184.
Pteropod and globigerina ooze, i. 264.
Pteropod ooze, depth at which it appears, i. 50, 147.
　depth at which found, i. 265.
　in Straits of Florida, i. 283.
Pteropod shells, in globigerina ooze, i. 283.
　decay and solution of, i. 283.
　Murray on solution of, i. 283.
Pteropod silt in Windward Passage, i. 235.
Pteropods, i. 190.
Pterotrachea, i. 191.
Ptychogena lactea, ii. 128.
Pulvinulina auricula, ii. 169.
Pulvinulina Menardii, ii. 169.
Pumice, decomposition of, i. 267.
　floated out to sea, i. 267.
　taken by tow-net, i. 267.
Pycnogonidæ, eyes of, ii. 39.
　Report on, by E. B. Wilson, i. xxi.
Pycnogonids, ii. 40.
Pylocheles Agassizii, ii. 40.
Pyrosoma, i. 187.
Pyrosomæ, i. 198.

Quoy and Gaimard on limit of reef-building corals, i. 74.

Radiaster elegans, ii. 104.
Radiella sol, ii. 170.
Radiolarian earth of Barbados, ii. 157.
Radiolarian ooze, i. 266.
Radiolarians, Johannes Müller on yellow cells of, i. 213.
    E. Haeckel on yellow cells of, i. 213.
    siliceous species of, ii. 157.
Raised terraces of Caribbean district, ii. 10.
Ramsay on geological time, i. 139.
Range of species, ii. 12.
Rawson, Rawson W., ii. 123.
Reade, M., on chemical denudation, i. 128.
Record of dredging and trawling, i. 32.
Red clay deposit, i. 267.
    basin of, in Eastern Atlantic, i. 203.
    basin of, in South Atlantic, i. 203.
    discovery of, by "Challenger," i. 202.
    in Western Atlantic, i. 202.
    slow rate of accumulation of, i. 268.
Redonda, island of, i. xvi.
Regadella phœnix, ii. 172, 173.
Regalecus Jonesii, ii. 28.
Regnard and Certes on decay at great depths, i. 203.
Regnard on effect of pressure, i. 305, 314.
Rein on undermining of Bermuda, i. 87.
Renard on St. Paul's Rocks, i. 127.
Rennell, i. 247.
Reophax scorpiurus, ii. 163.
Reptiles, distribution of, by driftwood, i. 122.
    relation of West Indian, i. 115.
Retepora reticulata, ii. 82.
Reuss on tertiary corals, ii. 18.
Reynolds, E. L., i. viii, 32.
Rhabdammina abyssorum, ii. 162, 163.
Rhabdammina linearis, ii. 163.
Rhabdocidaris, radioles of, ii. 90.
Rhabdoliths, i. 209.
Rhabdospheres, i. 209.
Rhamphobrachium Agassizii, ii. 55.
Rhizochalina, ii. 177.
Rhizocrinus, dredged by Pourtalès, i. 285.
    discovery of, i. 43.
    fields of, ii. 118.
    range of, ii. 15.
Rhizocrinus lofotensis, ii. 120, 121.
Rhizocrinus Rawsoni, ii. 121.
Rhizopod earth of Barbados, i. 79.
Rhizopods, ii. 157.
    association of arenaceous and siliceous, ii. 158.
    calcareous, living on bottom, ii. 159.
    Goës on Caribbean, ii. 157.
Rhizophysa, i. 185.

Rhizophysidæ, ii. 183.
Rhizostomæ, live on bottom, ii. 130.
Rhizotrochus fragilis, ii. 151.
Rhombodichthys, i. 175.
Rhynchopygus caribæarum, ii. 97.
Ribbon fishes, ii. 28.
Ridge, between Atlantic and Arctic oceans, i. 242.
    from Santa Cruz to Porto Rico, i. 98, 112.
Ringgold, i. 48.
Ringicula leptocheila, ii. 69, 70.
Ringiculidæ, ii. 69.
Rocinela oculata, ii. 48.
Rocky banks off the Carolinas, i. 276.
Rodgers, i. 48.
Roebling and Sons, wire rope of, i. 29.
Rogers, H. D., on character of Florida, i. 67.
Rogers, W. R., i. 47.
Rondelet's Medusa head, ii. 113.
Ross, James, i. 39.
    Jas. C., i. 42, 44.
Rosses on deep-sea temperatures, i. 46.
Rugosa, Ludwig on, ii. 154.
    Milne-Edwards and Haime on, ii. 154.
Ryder on organs of lateral line, ii. 36.

Saba Bank, i. xv.
Saba, island of. i. xv.
Sabinea princeps, ii. 45.
Sagartia abyssicola, ii. 147.
Sagitta, i. 193.
Sagrina dimorpha, ii. 166.
Saintes, the, i. xvi.
Salenia, i. 158; ii. 90.
    apical system of, ii. 91.
    suranal plates of, ii. 91.
    young, ii. 92.
Salenia Pattersoni, ii. 90, 91.
Salenia varispina, ii. 90, 91.
Salinity of sea water, i. 300.
Salpa, i. 187.
    large solitary form and chain, i. 189.
    masses of, i. 171.
Salpa Caboti, i. 190.
Salpæ, sudden appearance of, i. 190.
Salts, distribution of by animals, i. 153.
Sander Rang on Holopus, ii. 123.
Sandy bottom of East Coast, i. 285.
Sandy deposits, where found, i. 270.
Santa Cruz, "Albatross" on ridge between it and Porto Rico. i. 98.
    connection with Porto Rico and St. Thomas, i. 112.
    island of, i. xiv.

INDEX. 215

Santa Cruz mollusks, affinities of the, i. 112.
Saporta on formation of earth's crust, i. 140.
Sapphirina, i. 171.
Sarasin and Fol on penetration of light, i. 305, 306.
Sarasin and Soret on penetration of light, i. 306.
Sargasso Sea, i. 209.
   Columbus on, i. 213.
   "Talisman" Expedition on, i. 211.
Sargassum, i. 121, 210.
   animals inhabiting the, i. 212.
   animals living upon, i. 213.
   J. R. Bartlett on, i. 211.
   Alph. Milne-Edwards on, i. 212.
   Moseley on fructification of, i. 212.
   north of Cape Hatteras, i. 211.
   reproductive organs of, i. 212.
Sargassum bank, origin of, i. 213.
Sargassum fields, off Porto Rico and San Domingo, i. 211.
Sars, G. O., i. 44, 204.
   deep dredging off Norway, i. 42.
   on Rhizocrinus, i. 285.
Sars, M., animals from great depths, i. 42.
   on Rhizocrinus lofotensis, ii. 120.
Scæorhynchus armatus, ii. 50, 51.
Scalpellum regium, ii. 50.
Scammon, J. Young, ii. xi.
Schizaster, fascioles of, ii. 98.
Schizopods, large size of, ii. 48.
Schmidtia aulopora, ii. 178.
Schmidt, O., Report on Sponges, i. xxi.
Schultze, Max, i. 35.
   on chlorophyll in planarians, i. 213.
Scombroids, ii. 28.
Scopelidæ, ii. 33.
Scopelus, ii. 24.
   luminosity of, ii. 33.
   Mülleri, ii. 33.
Scoresby on deep-sea temperatures, i. 40.
Scorpæna, ii. 29.
Sculpins, ii. 29.
Scutellæ, absence of in deep water, ii. 97.
Sea-urchins, ii. 88.
   A. Agassiz, Report on, i. xxi.
   geographical range of, ii. 16.
   known previous to "Blake" Expedition, ii. 4.
   oldest known, ii. 94.
Sea water, air in, i. 207.
   analysis of, i. 21.
   Buchanan, J. Y., chemistry of, i. 23.
   Buchanan, J. Y., on oxygen in, i. 295.
   Dittmar composition of, i. 296.
   Dittmar report on samples of, i. 200.

Sea water, Dittmar on solvent action of, i. 283.
   elements in solution in, i. 206.
   Jacobsen on gaseous elements of, i. 294.
   Murray on carbonate of lime in, i. 65.
   Murray on solvent action of, i. 65.
   organic matter in, i. 205.
   specific gravity of, i. 20.
Secchi, on penetration of colors, i. 305.
Sedimentary rocks, thickness of, i. 130.
Seeds, transportation of, i. 180.
Seguenza, on deep-sea fossil corals, ii. 19.
   on deep-sea pliocene, i. 145.
   on tertiary corals, ii. 18.
Selachians, deep-sea, ii. 36.
   Garman, S., Report on, i. xxi.
Semper, i. 76.
   on coral reefs, i. 76.
Serpulæ, masses of, i. 83.
Serpulidæ, at great depth, ii. 57.
   bathymetrical range of, ii. 57.
Setidium obtectum, ii. 176.
Sharks in Lake Nicaragua, i. 153.
Sharks and whales, remains of, on bottom, i. 268.
Sharks' teeth, dredged by "Challenger," i. 145.
   found by "Challenger," i. 276.
   in concretions, i. 276.
Sharples, analysis of corals, i. 62, 148.
   analysis of rock of Pourtalès Plateau, i. 148.
Sharrer, W. O., i. viii, 32.
"Shearwater," cruise of, i. 40.
Sheaves for dredging, i. 33.
Shore lines and hydrographic basins, i. 100.
Siemens, C. W., bathometer, i. 6.
   electrical thermometer, i. 17.
Sigsbee, C. D., i. viii, 32.
   accumulator, i. 6.
   collecting cylinder, i. 36, 200.
   deep-sea sounding and dredging, i. 51.
   detacher, i. 3.
   exploration of Gulf of Mexico, i. 50.
   first dredging season in Gulf of Mexico, i. 37.
   gravitating trap, i. 36.
   on movements of Pentacrinus, ii. 119.
   Pentacrinus ground, ii. 6.
   sounding machine, i. 6.
Sigsbee Deep, i. 102.
Sigsbee, L. P., i. x, 32.
Sigsbeia, ii. 114.
Sigsbeia murrhina, ii. 5.
Silica, carried out to sea, i. 150.
   supply of, due to deep-sea sponges, i. 149.

Siliceous bottom deposits of Caribbean, i. 280.
Siliceous sponges, abundance of, i. 149.
Siliquaria modesta, ii. 71.
Silt, accumulation of, off Hatteras, i. 202.
    deposition of, to south of Windward Passage, i. 202.
    deposits of, i. 141.
    precipitation of, i. 138.
    transfer of masses of, i. 131.
Siphonophores, ii. 132.
    bathymetrical range of, i. 185.
Skates, deep-water types of, i. 193.
Slime, organic, at great depths, i. 203.
Smith, E. A., on geology of Florida, i. 110.
Smith and Hilgard on the Florida backbone, i. 61.
Smith, S. I., on range of crustacea, ii. 16.
    Report on Crustacea, i. xxi.
Smitt, i. 42.
    on range of bryozoa, ii. 15.
    Report on Bryozoa, i. xxi.
Snappers, ii. 28.
Solaster endeca, ii. 103.
Solenodon, i. 114.
Solid matter, held in solution, i. 128.
Solution of shells in deep water, i. 147.
Sombrero, island of, i. xix.
Sorosphera confusa, ii. 162.
Sounding, accumulator used in, i. 8.
    deepest, by Brownson, i. 50, 238.
    deepest, made, i. 97.
    error in, i. 12.
    in currents, i. 1.
    line for, i. 2.
    records of, i. 13.
    reel for, i. 7.
    reeling in pulley for, i. 8.
    register for, i. 8.
    time occupied in, i. 12, 13.
Soundings, along course of Gulf Stream, Bailey on character of, i. 272.
    along New England by U. S. Fish Commission, i. 273.
    extraordinary, i. 1.
    from Havana to Bahia Honda, i. ix.
    from Havana to Sand Key, i. ix.
Soundings: horizontal distance between the one hundred and the fifteen hundred fathom line, i. 101.
    one hundred fathom line, course of, i. 93.
    one hundred fathom line, absence of south of Hatteras, i. 135.
    one hundred fathom line, George's Bank to Hatteras, i. 93.
    one hundred fathom line south of Hatteras, i. 93.

Soundings: one hundred fathom line of West India Islands, i. 111.
    five hundred fathom line of West India Islands, i. 111, 112.
    off Montauk Point, Bailey on, i. 272.
    one thousand fathom line, i. 96.
    two thousand fathom line, i. 96.
    two thousand fathom line north of Cape Cañaveral, i. 96.
    twenty-five hundred fathom line, i. 97.
Sounding wire, weight of, i. 11.
Sounding with wire, i. 2.
    advantage of, i. 11.
    by "Blake," i. 50.
    Thomson on, i. 4.
South America, before the tertiary period, i. 116.
    disconnected from North America, i. 116.
South American fauna and flora, relations of, i. 109.
Southern equatorial current, i. 243.
Southern fauna, northern extension of, i. 110.
Species, localization of, i. 168.
Specific gravity, of North Atlantic, i. 298.
    of ocean water, i. 298.
    of rocks from great depths, i. 287.
    of sea water of American coast, i. 299.
Sphærozoum, i. 195.
Sphargis, i. 183.
Spirialis, i. 205.
Spirula, ii. 61.
    shell of, i. 160.
Spitzbergen, rising of coast of, i. 120.
Sponges, ii. 170.
    color of deep-sea, i. 312.
    distribution of, ii. 17.
    Haeckel on individuality of, ii. 170.
    in shallow waters of cretaceous, i. 149.
    of the "Blake," Schmidt on the, ii. 170.
    on West Bank of Florida, i. 149.
    Report on, by O. Schmidt, i. xxi.
    Schmidt on individuality of, ii. 170.
    siliceous, absent on east coast of U. S., ii. 170.
Sponge sarcode, i. 149.
Squids, giant species of, i. 191.
St. Eustatius, island of, i. xv.
St. Kitts, island of, i. xv.
St. Lucia, island of, i. xvii.
St. Paul's Rocks, i. 127.
St. Pierre, i. xvi.
St. Thomas, island of, i. xiv.
St. Vincent, island of, i. xviii.

INDEX. 217

Stalked crinoids, ii. 116.
Starfishes, ii. 102.
   bathymetrical range of, ii. 107.
   known previous to "Blake" Expedition, ii. 4.
   prominent deep-sea families, ii. 102.
   Report on, by E. Perrier, i. xxi.
Staurophora, i. 177.
Stauroteuthis, i. 191.
Steindachneria, ii. 26.
Stellwagen cup, samples brought up by, i. 261.
Stenocyathus vermiformis, ii. 148.
Stenoteuthis, i. 191.
Stenoteuthis Bartrami, ii. 58.
Stephanomia, i. 184.
Stephanotrochus diadema, ii. 149, 150.
Sternoptyx diaphana, ii. 22.
Sthenelais simplex, ii. 54.
Stichopus natans, ii. 85.
Stimpson, i. 43.
Straits of Florida, warm current from, i. 241.
Studer, on coral reefs, i. 76.
   on deep-sea siphonophores, i. 184.
Stylaster filogranus, ii. 139, 140.
Stylasteridæ, ii. 138.
Stylifer, ii. 64.
   Stimpson on, parasitic of annelids, ii. 64.
Styliola, i. 187, 205.
Stylodactylus, ii. 46.
Stylorhiza stipitata, ii. 177.
Submarine banks, discovered by "Challenger," i. 64.
   discovered by "Tuscarora," i. 64.
Submarine cables, i. 2.
Submarine deposits, i. 260.
   John Murray, Report on, i. xxi.
Submarine disturbances, i. 104.
Submarine landscapes, i. 103.
Submarine plateaux, formation of, i. 77.
Submarine ridges, i. 245.
Submarine scenery, monotony of, i. 106.
Submarine slopes, steepness of, i. 102.
Sulcastrella clausa, ii. 176.
Sulphate of lime, amorphous, i. 149.
Surface algæ, field of, i. 313.
Surface animals in bottom deposits, i. 285.
Surface fauna, A. Agassiz, Report on, i. xxi.
Swordfish, sounding of, ii. 24.
Syllis, phosphorescent species of, i. 199.
Symbiosis, i. 215.
   De Bary on, i. 214.
Synaphobranchus pinnatus, ii. 34, 35.
Sysconus infelix, ii. 48.

Tactile organs of deep-sea fishes, ii. 22.

"Talisman," Expedition of the, i. 40.
Tangle bar, i. 25.
Tanner on ridge between Santa Cruz and Porto Rico, i. 222.
Tasmania, relation to Australia, i. 125.
Telegraph cable, animals on, i. 40.
Temnechinus maculatus, ii. 92.
Temperature at one thousand fathoms, i. 248.
Temperature, belt of falling, i. 301.
   belt of uniform, i. 301.
   between Bahamas and Bermudas, i. 237.
   condition of, below five hundred fathoms, i. 302.
   differences of, over extensive areas, i. 303.
   effect upon fauna by change of, i. 119.
   highest, to which man is subject, i. 301.
   increase of, in interior of the earth, i. 303.
   in Florida Straits, i. 246.
   lowest, found by "Challenger," i. 246.
   lowest, found by "Porcupine," i. 245.
   lowest, to which man is subject, i. 301.
   of bottom of ocean, Gardner on, i. 132.
   of eastern and western continental shores, i. 244.
   of ocean, Foster on, i. 40.
   of sea-water of greatest density, i. 248.
   range of, for marine animals, i. 301.
   seasonal differences of, i. 246.
   uniform in deep water, i. 164.
Temperature and light in the tropics, i. 164.
Temperatures, of the Caribbean, i. 217.
   of the Gulf of Mexico, i. 217.
   of the Western Atlantic, i. 217.
   off Barbados, i. 227.
   off leeside Windward Islands, i. 228.
   off Salines Point, Grenada, i. 228.
   variable belt of, i. 247.
Temperature sections:
   across Mona Passage, i. 223.
   across Windward Passage, i. 224.
   across Yucatan Channel, i. 219, 230.
   by "Albatross" in Caribbean, i. 217.
   from Cape Florida to Gun Key, i. 232.
   from Dominica to Martinique, i. 222.
   from Halifax to Bermuda, i. 244.
   from Jamaica to San Domingo, i. 225.
   from Jupiter Inlet to Memory Rock, i. 232.
   from Madeira to Tristan da Cunha, i. 242.
   from Martinique to St. Lucia, i. 222.
   from Mexico to Florida, i. 231.
   from Pernambuco to Fernando Noronha, i. 227.
   from Santiago de Cuba to Jamaica, i. 226.

Temperature sections:
  from Sombrero to Virgin Islands, i. 219.
  from Sombrero to Virgin Gorda, i. 221.
  from St. Lucia to St. Vincent, i. 223.
  from St. Thomas to Bermudas, i. 220.
  from St. Thomas to Ham's Bluff, i. 221.
  from Teneriffe to Sombrero, i. 244.
  from Tortugas to Cuba, i. 231.
  from Tortugas to Yucatan, i. 230.
  from Vera Cruz to Galveston, i. 231.
  from Yucatan Bank to Louisiana, i. 231.
  from Yucatan to Santa Rosa, i. 231.
  in Gulf of Mexico, i. 230.
  in Gulf of Mexico, Sigsbee on, i. 217.
  of Caribbean and east coast of U. S., Bartlett on, i. 217.
  off Cape Cañaveral, i. 233.
  off Cape May, i. 239.
  off Charleston, S. C., i. 233.
  off Hatteras, i. 239.
  off the middle states, i. 238.
  off Montauk, i. 239.
  off New England, i. 238, 239.
  off Sombrero, i. 220.
  off St. Simon's Island, Ga., i. 233.
Terebellidæ, bathymetrical range of, ii. 57.
Terebratula caput serpentis, ii. 77.
Terebratula cubensis, ii. 76.
Terebratulina Cailleti, ii. 76, 77.
Terrestrial folds, developed by soundings, i. 125.
Terrigenous deposits, i. 263.
Tessadroma borenle, ii. 81.
Tetractinellidæ, ii. 177.
Textularia sagittula, ii. 164.
Textularia trochus, ii. 164, 165.
Thalassia, i. 82.
Thalassicolæ, i. 195.
Thalassography, i. 2.
Thecocyathus cylindraceus, ii. 149.
Thecopsammia socialis, ii. 153.
Thecopsammin tintinnabulum, ii. 152.
Théel, H., Report on Holothurians, i. xxi.
  on range of holothurians, ii. 15.
Thermic theory of Leonardo da Vinci, i. 249.
Thomson, C. Wyville, i. xx, 285.
  on Atlantic and Pacific gulfs, i. 242.
  on concentration and precipitation, i. 300.
  on stem of Pentacrinus, ii. 119.
Thomson Deep, i. 106.
Thomson, J. V., on young Comatula, ii. 116.
Thomson, Wm., sounding machine of, i. 4.
Thurammina papillata, ii. 164.
Tiedemannia, i. 187.
Tile fish, ii. 29.

Tile fish, destruction of, i. 120.
Tima, i. 177.
Tindaria cytherea, ii. 72.
Tisiphonia fenestrata, ii. 177.
Tizard, i. 44.
Tobago, island of, i. xix.
Tomopteris, i. 193.
Torell, collections of, noticed by Keferstein, i. 42.
Tortugas, action of the wind on the reef of, i. 86.
  bank west of, i. 88.
  broken ground of, i. 86.
  description of the, i. 80.
  examination of, i. 57.
  fauna and flora of, i. 90.
  formation of, i. 61.
  knoll of, i. 59, 89.
  line from, to Yucatan Bank, i. ix.
  line north of the, i. ix.
  physiognomy of the coral reefs of, i. 82.
  sections across the, i. 80.
  visit to, i. x.
Tow-net, i. 35,
  of Palumbo, i. 37.
Trachynema, i. 183.
Tradewinds and oceanic currents, i. 255.
Transportation of land shells and saurians, i. 291.
"Travailleur," Expedition of the, i. 40.
Trawl, modification of, i. 26.
  used by "Blake," i. 48.
  used by "Talisman," i. 48.
  used by U. S. Fish Commission, i. 48.
Trawl weights, i. 29.
Trawling, time of, i. 27.
  deepest of "Blake," i. 50.
Tremaulidium geminum, ii. 176.
Trichiuridæ, ii. 28.
Trichodesmium erythræum, i. 208.
Triforis longissimus, ii. 71.
Trigonocidaris albida, ii. 92.
Trinidad, island of, i. xix.
"Triton," Expedition of the, i. 44.
Trochidæ, ii. 67.
Trochostoma arcticum, ii. 85, 86.
Truncatulina adunca, i. 271.
Truncatulina Ungeriana, ii. 169.
Tubularians, littoral, ii. 136.
Turbot of New England, ii. 23.
"Tuscarora," soundings by, i. 200.
Typhis longicornis, ii. 70.

Udotea, i. 82.
Umbellula, collected by Adrians, i. 40.
Umbellula Güntheri, ii. 142, 143.

United States Fish Commission, i. xx.
Upper Gault, depth of sea of, i. 148.
Urechinus naresianus, ii. 101.

"Valorous," Expedition of the, i. 43.
Valvulina triangularis, ii. 165.
Vegetable parasites, Wedl and Kölliker on, i. 288.
Velella, i. 180, 183.
Ventriculites, Thomson, C. Wyville, on, ii. 170.
Vermetus erectus, ii. 71.
Verrill, A. E., Report on Anthozoa, i. xxi.
  on "Blake" anthozoa, ii. 142.
  on "Blake" cephalopods, ii. 58.
  on color of light at great depths, i. 305.
  on fish remains, i. 145.
  on floating beach sand, i. 274.
  on pliocene submarine fossils, i. 273.
  on primitive types of actiniæ, ii. 146.
  on protective phosphorescence, i. 308.
Verruca incerta, ii. 50.
Vertebrate bones, scarcity of, i. 144.
Verticordia elegantissima, ii. 74.
Verticordia perversa, ii. 74.
Vesicomya pilula, ii. 74.
Vesicomya venusta, ii. 74, 75.
Vetulina stalactites, ii. 175.
Vicksburg limestone. i. 61.
Vincularia abyssicola, ii. 80.
Virgin Islands, i. 105.
  land shells of, i. 116.
  sink off, i. 104.
  submarine banks of the, i. 111.
Volcanic bottom deposits, i. 290.
Volcanic regions, topography of bottom of, i. 104.
Volcanic shore deposits, i. 289.
Volcanoes of West Indies, age of the, i. 109.
Voluta, bathymetrical range of. i. 169.
Von Otter, i. 42; ii. 142.
"Vöringen," expedition of the, i. 44.
Vorticellidæ, associated with other animals, i. 215.

Waldheimia floridana, ii. 76.
Wallace, A. R., i. 109.
  on age of continents and oceanic basins, i. 127.
  on Antillean continent, i. 116.
Wallich, on formation of flints, i. 143.
  on migrations of marine animals, i. 104.
  on protoplasmic deep-sea layer, i. 150.
  on silex nodules, i. 150.
  report on "Bulldog" Expedition, i. 44, 45.

Wallis, i. 32.
"Washington," Expedition of the, i. 41.
Water, disintegrating action of warm, i. 290.
Water cup, i. 21.
  of Sigsbee, i. 21.
  of Tornöe and Wille, i. 294.
West India Islands, appearance of, i. xiii.
  elevated reefs of, i. 79.
  eruptive rocks of, i. 110.
  history of, i. 111.
  limestone of, i. 64.
  southern slope of, i. 105.
  time of elevation of, i. 113.
West Indian bird fauna, i. 114.
West Indian cretaceous rocks, i. 110.
West Indian deep-sea fauna, richness of, ii. 3.
West Indian fauna, ii. 1.
  immigration into, of Atlantic types, i. 158.
  transition from old fauna, i. 160.
  variety and richness of, i. 91.
West Indian fauna and flora, origin of, i. 110.
  relations of, i. 109.
West Indian marine animals, northern extension of, i. 119.
West Indian miocene rocks, i. 110.
West Indian reptiles, relation of, i. 115.
West Indian specific forms, i. 116.
West Indian submarine plateau, i. 113.
West Indian types, northern extension of, i. 118.
West Indies, Duncan on fossil corals of, i. 101.
  fossiliferous rocks of, Cleve on, i. 109.
  land tortoise of, i. 115.
  mollusk fauna of, i. 114.
Western Atlantic, bird's-eye view of, i. 105.
Western Caribbean, topography of, i. 100.
Western North Atlantic, warm water of, i. 243.
White chalk, a deep-sea deposit, i. 146.
  composition of, i. 291.
  off Nuevitas, i. 280.
Wild, J. J., i. 246.
  Thalassa, i. 244.
Willemoesiæ, ii. 42.
Willemoes-Suhm on luminosity of Scopelus, ii. 33.
Wilson, E. B., ii. 49.
  Report on Pygnogonidæ, i. xxi.
Winds, effect of, in depth, i. 255.
  frictional effect of, i. 247.
Windward Islands, bottom of plateau of, ii. 158.

Windward Islands, limestone terraces of, i. 64.
Windward Passage, depth of, i. 100.
   Bartlett on current passing over ridge of, i. 292.
Winn, i. x.
Winsor, Justin, i. xxi.
Wire rope, for dredging, i. 28.
   kinking of, i. 30.
   paying out of, i. 33.
   reel for, i. 31.
   telephoning of, i. 30.
Worms, ii. 52.
Wright, i. x.
Wyman, Jeffries, i. 115.

Xylopagurus rectus, ii. 40.

Yucatan Bank, i. 141.
   depth at foot of, i. 101.
   formation of, i. 69.
   north slope of, i. ix.
   Vicksburg limestone of, i. 69.
Yucatan Channel, depth of, i. 101.
Yucatan, limestone backbone of, i. 110.
   limestone plateau of, i. 72, 122.

Zanclea, i. 183.
Zittel on teeth of Conoclypus, ii. 90.
Zoeppritz on friction of particles of water, i. 255.
Zoögeographical divisions, i. 264.
Zoroaster Ackleyi, ii. 105.
Zoroaster Sigsbeei, ii. 105.
Zygodactyla, i. 177.

www.ingramcontent.com/pod-product-compliance
Lightning Source LLC
Chambersburg PA
CBHW030018240426
43672CB00007B/1007